83

Advances in Biochemical Engineering/Biotechnology

Series Editor: T. Scheper

Springer-Verlag Berlin Heidelberg GmbH

Proteomics of Microorganisms
Fundamental Aspects and Application

Volume Editors: M. Hecker · S. Müllner

With contributions by
D.J. Cahill, P. Cash, S.J. Cordwell, M. Hecker, H.E. Meyer, M. Mreyen, S. Müllner, E. Nordhoff, A.S. Nouwens, W. Schubert, A. Sickmann, R.A. VanBogelen, B.J. Walsh

Springer

Advances in Biochemical Engineering/Biotechnology reviews actual trends in modern biotechnology. Its aim is to cover all aspects of this interdisciplinary technology where knowledge, methods and expertise are required for chemistry, biochemistry, microbiology, genetics, chemical engineering and computer science. Special volumes are dedicated to selected topics which focus on new biotechnological products and new processes for their synthesis and purification. They give the state-of-the-art of a topic in a comprehensive way thus being a valuable source for the next 3–5 years. It also discusses new discoveries and applications.

In general, special volumes are edited by well known guest editors. The series editor and publisher will however always be pleased to receive suggestions and supplementary information. Manuscripts are accepted in English.

In references Advances in Biochemical Engineering/Biotechnology is abbreviated as *Adv Biochem Engin/Biotechnol* as a journal.

Visit the ABE home page at
http://www.springerlink.com/series/abe/

ISSN 0724-6145
ISBN 978-3-642-05607-9 ISBN 978-3-540-36459-7 (eBook)
DOI 10.1007/978-3-540-36459-7

Library of Congress Catalog Card Number 72-152360

http://www.springer.de

Originally published by Springer-Verlag Berlin Heidelberg New York in 2003

Softcover reprint of the hardcover 1st edition 2003

Typesetting: Fotosatz-Service Köhler GmbH, Würzburg
Cover: KünkelLopka GmbH, Heidelberg/design & production, Heidelberg

Printed on acid-free paper 02/3111mh – 5 4 3 2 1 SPIN 11396505

Advances in Biochemical Engineering/Biotechnology also Available Electronically

For all customers with a standing order for Advances in Biochemical Engineering/Biotechnology we offer the electronic form via SpringerLink free of charge. Please contact your librarian who can receive a password for free access to the full articles. By registration at:

http://www.springer.de/series/abe/reg_form.htm

If you do not have a standard order you can nevertheless browse through the table of contents of the volumes and the abstracts of each article at:

http://www.springerlink.com/series/abe/

There you will find also information about the

- Editorial Board
- Aims and Scope
- Instructions for Authors

Attention all Users of the Springer Handbook of Enzymes

Information on this handbook can be found on the internet at
http://www.springer.de/enzymes/

A complete list of all enzyme entries either as an alphabetical Name Index or as the EC-Number Index is available at the above mentioned URL. You can download and print them free of charge.

A complete list of all synonyms (more than 25,000 entries) used for the enyzmes is available in print form (ISBN 3-540-41830-X).

Save 15 %

We recommend a standing order for the series to ensure you automatically receive all volumes and all supplements and save 15 % on the list price.

Preface

Starting with the discovery of penicillin, other antibiotics, and insulin, the quest for understanding and use of biological systems, i.e., microorganisms and animal tissue, for the production of value products has lead to a dramatic increase in microbiological and bioengineering research in the last decades. Chemical and pharmaceutical companies quickly realized the huge commercial potential of these bioproducts and have spent millions of US dollars on R&D as well as on a build-up of production facilities. Although there was limited knowledge about the cell's molecular mechanisms, which are the basis for the formation of the desired products, products from fermentation and extraction of biological matrices were a success right from the start. R&D projects within industry and academia on the continuous improvement of production processes, especially microbial productivity and down stream processing, allowed a fast return of investment and secured competitiveness in the market. Whereas the focus of such research projects was mainly on the discovery of strains with higher productivity for the product of interest, e.g., antibiotics, a lot of expertise and knowledge was generated allowing the use of biotechnological products and processes outside the pharmaceutical arena. The tremendous increase in knowledge and the technological developments in microbial genetics where driven by these research projects and, accompanied with the advancements in nucleotide chemistry leading to a much better understanding of intracellular processes, served as a basis for modern molecular biology and recombinant biotechnology.

Since the late 1970s, the success of recombinant pharmaceutical products was the major impetus for the developments in academic research and industrial biotechnology. New interdisciplinary research platforms were created and introduced into the scientific community, e.g. genomics, bioinformatics, proteomics, and, just recently, systems biology. At the same time, biotechnology has matured into a solid and highly profitable business area with impressive growth rates and turnover. Biotech products and products developed by using biotechnology have brought innovation to a great number of market segments ranging from forensics, analytics, detection of fakes, speciality chemicals, detergents and cleansers, cosmetics, consumer products, agriculture, diagnostics, food and feed, up to pharmaceuticals and medicine. All these market innovations, however, were heavily dependent on the targeted use of synergistic combinations of already existing as well as on the development of new technologies.

In this volume the editors have focused on one of the presently most exciting of these new technologies – proteomics. Whereas the pharmaceutical industry expected genomics and proteomics to deliver proprietary and validated (new) drug targets faster and thereby lead to a shortage in overall development time, it was the application of proteomics to microbiology and symbiosis, classical biotechnology and fermentation processes which has already generated valuable as well as applicable results for the improvement of industrial biotechnology. The understanding of metabolic (protein) networks within an industrially used cell is of clear importance for productivity of the organism and the whole process. Since, cell metabolism is influenced in many ways by the fermentation conditions, e.g., aeration, pH, media, bioreactor volume, temperature, mechanical stress, cell density, feedback effects of the product etc., proteomics is presently the method of choice for identifying functional coregulated and cooperating protein networks. Excellent experimental results with respect to the elucidation of relevant regulatory networks in microorganisms were obtained in academic as well as industrial research, and the generated knowledge could be successfully applied. Further advancements in this area are strongly dependent however, on new and more sophisticated technologies. Especially the study of cooperating protein networks and metabolic fluxes needs technologies with higher precision on the sub-microliter scale.

In order to cover the whole range of aspects of the application of proteomics and some selected supporting technologies for microorganism- and biotechnological processes, it was crucial to select a group of contributing authors not only from academia but also from the pharma and biotech industries.

The present volume brings together results, opinions and suggestions of some of the world's leading experts in the field of proteomics and evaluates its impact on products and processes.

Greifswald/Leverkusen, April 2003 Michael Hecker and Stefan Müllner

Contents

Adv Biochem Engin/Biotechnol (2003) 83: 1–25
DOI 10.1007/b10944

The Impact of Proteomics on Products and Processes

Stefan Müllner

Senior Vice President Life Sciences, Fundamenta Capital AG, Bergische Landstrasse 67, 51375 Leverkusen, Germany. *E-mail: muellner@fundamenta.de*

Not much more than 15 years ago a handful of visionary scientists around the world suggested to sequence and analyze not only the human genome but also as many genomes as possible in order to compare DNA as well as to deduce protein sequences. By that means they expected to get an idea about the organization of life. However, after now having now sequenced the human genome and at least identified around 40,000 genes as coding regions, we are still left with the fundamental questions of how genes are regulated, and what is the rationale of genetic regulatory networks.

The basic knowledge and methodologies to elucidate functional regulatory networks of cells and organisms on the protein level had been around for much longer than DNA-based discovery tools. This was mainly due to the fact that proteins have to fulfill universal functions in nature and, unlike DNA polynucleotides, proteins differ not only in their amino acid sequences; they come in nearly all shapes and sizes and have all kinds of physical as well as chemical properties. They can be highly water soluble, e.g., serum and milk proteins, or nearly insoluble in any solvent, e.g., keratin and some other structural proteins. In addition, structure, function, as well as the respective stability of proteins inside and outside of a biological system, are individual features of any given polypeptide. On one hand, the individuality of proteins allows adaptation of any life form to the environment, and on the other it is still a real challenge for biotech R&D and production.

The present review is actually the first approach to evaluate and judge the achievements made by Applied Proteome Analysis and Proteomics over the last 27 years.

Keywords. Proteome, Proteomics, Insulin, Protein chemistry, Protein maps, Mass spectrometry, Fermentation, Bioengineering

1 Introduction

Not much more than 15 years ago a handful of visionary scientists around the world made the suggestion to sequence and analyze not only the human genome but also as many genomes as possible in order to compare DNA as well as protein sequences. By that means they expected to get an idea about the organization of life. However, after now having sequenced the human genome and at least identified around 40,000 genes as coding regions, we are still left with the fundamental questions of how genes are regulated, and what is the rationale of genetic regulatory networks.

The development of solid phase chemistry for biomolecules by Letsinger [1, 2] and Merrifield [3, 4] and the discovery and introduction of restriction enzymes by Paul Berg and others [5–8], were milestones which revolutionized our view of biological systems. The focus of scientific work was changed from macroscopic biological phenomena to the molecular level.

The major achievements in biotechnology of the last two decades were based on the discoveries in the DNA world – the characteristics of deoxyribonucleotides combined with codons, exons, and introns organized in operons and regulons as well as regulatory elements. It is now obvious that the information stored at the DNA level can code for a much higher number of proteins, optimized and specifically adapted to their cellular tasks by a combinatorial use of exons, i.e., differential mRNA splicing.

Furthermore, several possibilities for posttranslational modifications of proteins are already known, e.g., proteolytic processing, glycosylation, three different phosphorylation types, -*N*-terminal acylation and other modifications, farnesylation, glycosylphospatidyl inositol anchoring, etc., which have not yet been found to be genetically encoded and which tremendously inflate the actual number of proteins. And with increasing sensitivity of technologies in protein chemistry other relevant types of postranslational modifications will be discovered.

1.1 Gene Expression and Protein Degradation

The intracellular molecular mechanisms and signals controlling protein as well as RNA degradation are another area which is still not well understood, but controlled proteolysis of transiently needed regulatory proteins, digestion of incorrectly folded or damaged protein as well as mRNA degradation are essential for a viable cell and therefore of high relevance for the expression yields of the desired products. Posttranscriptional regulatory processes which control gene expression on the RNA level are more or less a white spot in the map of biology. The spliceosome, a multifunctional complex of small RNAs and RNA-binding proteins, controls correct production of the blueprints used for protein synthesis. It is obvious that the understanding of the processes which control gene expression after RNA synthesis will have a major impact on the future of biotechnology.

1.2 Receptor Signaling and Intracellular Signal Transduction

Complex reaction cascades between proteins can take place without any up or down regulation of a gene or expression of a gene product. Furthermore, specific intracellular transport processes are responsible for the compartmentation of cellular proteins and therefore the adaptation of cells to changes in the environment.

Since those protective responses have to occur immediately, an involvement of gene activation/expression is not favorable.

1.3 Carbohydrates and Lipids

The role of lipids and carbohydrates and how those molecules are involved in cellular processes is still one of the least understood areas in biochemistry. The same is true for posttranslational modification of proteins with carbohydrate and lipid moieties and their relevance to cellular compartmentation.

1.4 RNA – Information and Function

RNA serves as an intermediate or blueprint between the DNA-data pool – the genome – and protein function pool – the proteome; on the other hand some RNAs, so-called ribozymes, can catalyze chemical reactions like proteins can, or play important roles in ribosome or spliceosome structure and function as well as in other functional cellular complexes.

1.5 Synthesizing Life

Based on this knowledge, Szostak and Eigen developed a hypothesis that life started with the RNA world in which primordial cells lacking protein synthesis use RNA both as the repository of "genetic information" and as "enzymes" that catalyze metabolism [9, 10]. In view of the advances in directed evolution and membrane biophysics, Szostak more recently included phospholipid self-assembly and vesicular compartmentation of RNA molecules into his concept, and envisions the synthesis of simple living cells as an imaginable goal [11].

1.6 Proteomics – Basis for Advanced Biotechnology

However, even if the process of gene transcription in bacteria and higher organisms is by far the best understood area in biotechnology, the knowledge base is still weak and only a very few organisms can be considered as (relatively) well understood.

Over the last ten years, parallel as well as subsequent developments in totally different scientific areas – from Atomic force microscopy to relevance of Zinc in metallo proteinases – were made and created the new scientific field proteomics. Today, a direct linkage of 2-DE (2-dimensional polyacrylamide gel electrophoresis), micromethods in protein chemistry, mass spectrometry and biologic data bases is feasible.

The present volume is focused on the relevance of new technological developments in proteomics for microorganisms used in biotechnology, biotechnological products and processes. However, the major financial and technological driving force for the advancement of proteomics technologies is its huge potential to speed up pharmaceutical R&D.

2 Historical Aspects of Proteomics Development

Proteomics is a field, just as genomics is, rather than a closed and conceptually static body of knowledge and the understanding of the much more complex protein world and its linguistics will rely totally on the development of more sensitive and faster analytical nano-methods as well as new computational tools and algorithms.

It was in 1994 that Keith Williams and Mark Wilkins first coined the term proteome, basing it on the term genome, which is defined as the entirety of all genes and all not-protein-encoding DNA sequences of a given organism. According to their definition, proteome, then, refers to the entirety of all proteins that can be deduced from all the coding sequences of a genome of a particular organism. Due to the fact that no known organism needs and uses all gene products at the same time and in the same concentration, the term proteome today comprises the whole set of proteins directly linked to the metabolic state of a tissue or organism and precisely defined by descriptive experimental parameters.

2.1 Insulin – The Most Relevant Molecule for the Advancement of Protein Chemistry

Going back in history we will find that proteomics and proteome analysis are based to the largest extent on the achievements in protein sciences and physiology. The key driver for steady improvement of relevant technologies was and still is medicinal research and the pharmaceutical industry. In this respect, insulin, the peptide hormone which is produced by the pancreas and which is responsible for glucose homeostasis in all mammals, can be viewed as the molecule with the highest impact on this technological development. Insulin was discovered by Banting and Best in 1923. This discovery was made when protein chemistry was in an early embryonic stage but the specific features of this natural protein or peptide soon allowed its therapeutic use. Since then protein chemistry and protein analysis became an important area in the development of therapeutic protein drugs, blood plasma and serum products, as well as vaccines. However, it took more than 30 years to come up with the first protein sequence, which was again insulin. Paralleled by the progress made in nucleic acid chemistry, scientists realized that protein sequences are precisely encoded by DNA and are not random chains of amino acids. Starting with the early 1960s, evolving electrophoretic and chromatographic methodology allowed one for the first time to analyze and to separate complex protein mixtures from different biological sources. When the molecular biology revolution started in the 1970s, the first gene which was expressed in *Escherichia coli* (*E. coli*) bacteria was human insulin. Protein science was at that time just in its infancy. Nevertheless, a few years after scientists at Genentech showed that the production of human insulin in *E. coli* is possible, Eli Lilly decided to invest in this new technology and developed an industrial production process for recombinant human insulin. This can be regarded as a milestone in the history of recombinant biotechnology and the beginning of a new industrial sector, the biotech industry.

2.2 Regulatory Demand and Technology Development

By working of the stepwise improvement of this and other production processes for therapeutic proteins, a demand for all kind of new techniques evolved, and due to the success of biotech products major investments in this area were made. The fast developments in molecular biology, genetics, cell biology, microbiology and classical biotechnology were only possible by parallel improvements in protein chemistry, amino acid analysis, protein sequencing, enzymology, immunology, polymer chemistry, carbohydrate chemistry, LASER physics and physical chemistry, membrane technology, chromatographic methods, mass spectrometry and computer sciences.

Historically, the portfolio of technologies which are now commonly summarized as proteomics has evolved mainly in academic groups with clear focus on protein analysis. Research there was solely technology driven. Regulatory administrations like the FDA, however, put increasing pressure on drug companies involved in the development and marketing of therapeutic proteins and created by that means a huge demand for new technologies in protein chemistry. That is why the developments described here were paralleled by the efforts of pharmaceutical companies to get the approval of drug regulatory administrations for the marketing and genetically engineered therapeutic proteins and the steadily increasing demand for product quality and drug safety.

2.3 Impact on Exploratory Research

Besides a broad applicability of proteomics in pharma-related research, plenty of other important applications in all fields of biotechnology are possible. For instance, optimization of bacterial or fungal strains used in fermentation processes is feasible by subtractive protein pattern analysis and identification of metabolic key enzymes and regulatory proteins. The use of proteomics further allows not only the identification of new pharmacological targets for drug intervention, but also makes the characterization of tissue specific expression products in animals and plants feasible and allows the subsequent isolation of the respective genes as well as regulatory DNA sequences.

2.4 Pioneers and their Place in Time

Leonardo da Vinci invented the parachute more than 300 years before people started constructing and building airplanes. And with him a lot of engineers, scientists, and inventors from past to present had the sometimes demotivating experience that every idea has its time and some of those ideas and technologies are far ahead of their time.

And like Leonardo's parachute, the key technology for understanding the complex mixture of the cellular proteins – resolution two-dimensional sodium dodecylsulfate polyacrylamide gel electrophoresis (2-DE) – has been in existence for over 25 years for the subtractive protein pattern comparison of biological materials by means of high developed methods published independently by Klose [12] and O'Farrell [13] in 1975. Separation of hundreds up to 10,000 proteins by means of two-dimensional gel electrophoresis according to their protocols was already possible when no other supporting technologies for making use of the generated information was available.

Whereas O'Farrell in his original article already referred to the broad applicability "high resolution, sensitivity, and reproducibility make this technique a powerful analytical tool which could potentially find use in a wide range of investigations", Klose was fascinated by the fact that embryonic development of mice as well as genetically and teratologically connected problems can be studied by a single and comparatively simple experiment. Norman and Leigh Anderson, at that time still at Argonne National Laboratory working on analytical techniques for cell fractionation, especially zonal centrifuges, immediately realized the huge potential of 2-DE. This was long before the genomics age. Then 2-DE seemed to be the method of choice for studying biological complexity on a molecular level. After evaluating the possibilities of 2-DE and the development of improved protocols, equipment (ISO-Dalt system), and concepts, the Andersons conceptually designed in 1980 the Molecular Anatomy Program in order to allow the construction of the biological equivalent of the periodic table for man: the generation of the Human Protein Index, or as it would be called today, the Human Proteome Project.

2.5 Automation, Bottlenecks, and Robots

The methodology has improved over the years, but the possibilities for automation are still limited. Furthermore, the nature of proteins, which, unlike DNA, differ not only in size but also in their pI (isoelectric point), lipophilicity, and relative concentration or dynamic range, hampers the overall resolution, and still the present techniques are lacking the reproducibility, sensitivity, and selectivity required to study 100% of the proteins of a given cell or tissue.

Therefore, 2-DE is still not a common easy-to-use method and there are several reasons for this. First, well trained, experienced, and dedicated personnel are essential in every step of the laborious and cumbersome procedures to generate high performance two-dimensional protein gels – (a) precise definition of the biological problem and careful recording of all parameters describing it, (b) artifact minimizing, reliable and individually standardized protocol for sample preparation, (c) careful selection, storage and documentation on chemicals and solvents, especially water, used in all experiments, (d) standardized procedures for the first and second dimension, (e) awareness of limitations of all published and used protein staining protocols and reproducible adjustment to the given biological problem, (f) careful selection of data recording and storage tools as well as quantification algorithms, (g) use of the best suited and straight forward protein identification technologies, e.g. immunological identification by Western Blot, *N*-terminal sequencing of the full length protein blotted on a suitable membrane (PVDF of glass fiber), *N*-terminal sequencing of peptides after tryptic *in gel*-digestion of polyacrylamide protein spots (peptide fingerprints), mass spectrometric analysis of peptide fingerprints, (h) computerized comparison of experimental peptide masses with protein data bases.

Second, the polyacrylamide gel-based technologies do not allow automation of some crucial steps. Quantification is further heavily influenced by the staining procedures employed and can therefore only be relative. In addition, present software systems for computational handling of the generated data have been significantly improved in the last two years but still have dramatic limitations.

2.6 Reproducibility of Protein Maps

Reproducibility was the key issue of the early years of 2-DE after the first publications in 1975. In particular, interlaboratory comparison of 2-DE patterns was a real challenge until in 1983 Angelika Görg and Pier Giorgio Righetti [14] introduced the immobilines – defined acrylamide derivatives forming an immobilized pH gradient for the first dimension – marketed today by Amersham Pharmacia Biotech.

2.7 Micromethods in Protein Chemistry

Microanalysis of the protein spots of interest based on 2-DE gels was not possible before Ruedi Aebershold developed in 1985 a method for direct -*N*-terminal sequencing of peptides blotted to glass fiber membranes [15].

Whereas the possibility of protein transfer from polyacrylamide gels to chemically inert supports like glass fibers and PVDF-membranes dramatically changed the strategies in protein chemistry, there was still a need for a reliable and reproducible method for the generation of peptide fragments (maps) through direct enzymatic cleavage of protein spots within the polyacrylamide gel matrix. Lottspeich and Eckerskorn [16] published the first protocol for such a reliable micromethod in 1989.

At first this procedure was developed to provide access to a peptide mixture of for *N*-terminal sequencing and to circumvent the problem of a chemically blocked *N*-terminus which was, at that time, a common problem in the analysis of blotted proteins. Now it provides the basis for high grade automation in protein analysis. Especially with the increase of DNA and deduced protein sequences available in data bases it became more and more possible to identify proteins by just one or two isolated peptide sequences and to characterize the parent protein(s) function by homology.

2.8 PCR's Impact on Protein Chemistry

The advancements in protein chemistry in the 1980s were paralleled by the invention of the polymerase chain reaction (PCR). The broad applicability of this technology not only accelerated planned or ongoing DNA-sequencing projects. It was in addition the technological basis for all of the genome projects, and revolutionized existing methodologies and strategies in molecular biology, pharmaceutical and agricultural research, as well as diagnostics.

This was the starting point for the foundation of a growing number of genomics start-up companies which peaked in the mid 1990s and of large investments of all the major pharmaceutical firms into their R&D programs focused mainly on the comparison of differential gene expression in different tissues and metabolic states of organisms. The rationale of the pharma industry was to get fast access to unique and proprietary targets for the treatment of diseases. However, all the genomic information which can be generated by this approach has to be traced back to the exact functions of the identified genes. Since the relevance of a gene is reflected in the abundance and function of its gene product, further developments had to focus on proteins.

2.9 The Mass Spec Revolution

The impact of enzymatic peptide fingerprints on modern protein chemistry and the development of proteomics increased dramatically by the fast development of mass spectrometry paralleled by the exponential progress in computing power, and was propelled by a new strategy for fast protein identification. A milestone for employing this strategy was the development and application of mass spectrometric methods for the analysis of large biomolecules, especially proteins. Besides electrospray mass spectrometry, one of the major breakthroughs in this area was the development of matrix-assisted laser desorption ionization mass spectrometry or MALDI-MS by Hillenkamp and Karas in 1988 [17].

Having those new technologies at hand, fundamental work was done by John Yates at the University of Washington, who introduced the computerized comparison of peptide fingerprints and the calculated respective peptide masses generated by enzymatic cleavage in silico with the masses of peptide fingerprints from biological samples. By that means, fast, accurate and straightforward protein analysis and identification is feasible. The possibilities of the so-called Yates algorithm for the first time allowed thinking about high throughput in protein analysis. It also initiated an exponential growth in technological development. With the turn of the century, modern mass spectrometry has nearly replaced protein characterization by *N*-terminal sequencing. In addition, manufacturers of mass spectrometers have started their own business activities for the fast growing proteomics market or have formed alliances with companies of the electrophoresis and laboratory equipment business. The ultimate goal of these activities is to design and build an automated robotic system for proteome analysis.

2.10 The Ultimate Goal – Understanding the Proteins' Language

However, even the ability to generate excellent and reproducible protein patterns of a given cell type or tissue just allows an enumeration of the respective individual proteins present in the biological sample at a given time point, characterized by environmental or, more specifically, descriptive parameters. It is very important to generate this type of information and it can be already very helpful if the set of descriptive parameters allows a data reduction and information selection, but it provides no information about protein-protein interactions, the organization of protein complexes, the cellular compartmentation of proteins. It is further important to understand how proteins interact with DNA and RNA as well as with primary and secondary cellular metabolites. It is like the discovery of an ancient book in a foreign language where you have first to figure out the language, then you can learn the words, the writing, the grammar, you can start to read chapter by chapter, and after that you will start to think about the overall meaning of the book's content.

3 The Challenge – Applications of Proteomics in Life Sciences

3.1 The Proteome Reflects Living Dynamics

Proteome analysis as such is a very complex problem with every cell expressing between 4000 (bacteria) and 25,000 (mammalian liver) proteins in concentrations from just a few copies (transcription factors, regulatory proteins) up to 10^6 and more copies in a single cell. Besides other limitations of existing technologies in proteomics there is a hitherto unmet need for new technologies and strategies to deal with huge dynamic range of proteins within a cell.

The presence of specific proteins – not only regulatory proteins – within a cell, i.e., microorganism, tissue, organ, or whole organism, represents always a dynamic adaptation of a given metabolic state **A** reflected by a set of proteins or proteome **A** to changes in the descriptive parameters ($\mathbf{\Delta}_{\mathrm{Dp}}$) of the environment resulting in a metabolic state **B** reflected by a set of proteins or proteome **B**.

Microorganisms, for instance, use their receptors, chemical sensors, transporters, and ion channels to get constant access to the best nutrients in appropriate concentration. Exceeding a given threshold, perturbations of a biological system, e.g., a change in the nutrient source, oxygen or CO_2 concentration, toxins, drugs, xenobiotics, hormones, trauma, disease or stress always have influence on the metabolic state of cells in general, by that means affecting the proteomes of the cell types present in a tissue, organ or the whole organism.

3.2 The Normalome and the Need for Standards

Provided that mammals including man have around 200 morphologically clearly distinct cell types with a characteristic "normalome" (normalized proteome) defined only by sex and age of this particular organism, measurable changes in all of the 200 normalomes will occur through any perturbation.

Whereas the genome of a microorganism represents a huge but still countable number of DNA building blocks and therefore static body of information, an infinite number of proteomes is already possible. With higher organisms one has to cope not only with the comparison of a minimum of 200 normalomes but also with metabolically defined, development-stage-dependent, sexual, age-related proteomes, as well as proteomes created by targeted perturbation of a whole set of normalomes.

This way of looking at things, however, makes it finally evident that genome analysis and genomics are not the true bottom line. Every textbook of biology explains that proteins embody the functions within a cell and therefore stand for the active life, while DNA and RNA represent only plans. In other words, the genome is like a compact disc and the proteome is the music.

The genome is a constant for a given organism and comprises the complete program for this respective life form. However, as Leigh Anderson has pointed out, it's more to paella than the recipe, more to Bach than ink on paper, and more to society than its code of laws. Now, anyone can have access to the notes that encode life, but the conductor and the quality of the musicians in the orchestra, as well as the arrangements and instrumentation, are not known.

Therefore, defining life from the DNA level will end up in cacophony rather than harmony.

4
Applied Proteomics

4.1
Drug Mode-of-Action and Drug Targets

The relevance of proteomics for the pharmaceutical industry has been discussed over the last three years at a lot of scientific and commercial conferences and several visions since the pioneering paper by the Andersons in 1979 [18] have been proposed to apply the proteomics approach to catalogue all human proteins and to use this book on human molecular anatomy, the Human Protein Index, for target discovery and validation, drug mode-of-action (MOA) studies, and toxicological evaluation of chemicals and drugs [19–26]. In extensive series of in vivo studies of drug effects it was observed that proteins whose abundance or structure is strongly regulated by a drug or chemical are directly linked with the mechanism of drug action [27–29]. And by that means several potentially new pharmacological targets to be used in drug discovery programs were identified, but not all of them are suitable for industrial drug screening programs or can be used directly, especially in high-throughput screening (HTS) formats.

The unbiased multiparallel proteomics approach applied in target discovery and drug mode-of-action-studies normally delivers results which show characteristic changes of a whole set of proteins. Some of those might be directly linked to drug action. Most of them, however, are indirect or co-regulatory effects on protein abundance and activity. Therefore, the experiment has to be designed in such a way that a straightforward sieving process for unequivocally relevant targets or validated targets can be applied. Whereas genomics, and, due to close linkage to cellular function to a much lesser extent, proteomics, produce an overflow of possible, potential, and more or less relevant targets, the experimental pharmacologist always faces the acute problem of developing an experimental animal model in which he can clearly show the in vivo relevance of a respective target protein. Animal testing as such is the most expensive and time consuming step in preclinical development, and animal models are generally acknowledged by clinicians and drug regulatory administrations to be only few in number. In addition, the establishment and introduction of any new animal model needs time, patience, and money.

In order to be bought, used, and applied by the pharmaceutical industry, the best version of a validated target must fulfill the following criteria:

- Suitable for HTS, preferably a non-cofactor dependent enzyme
- Expression in *E. coli* should be possible, no-posttranslational modification with relevance for activity and stability
- Good stability under HTS-assay conditions
- Direct and proven correspondence between in vitro biochemical and in vivo pharmacological data
- An already existing animal model can be employed

In-house research programs as well as proteomics companies offering their services to big pharma firms have to meet these high class requirement standards. Presently, proteomics approaches allow the precise understanding of the MOA or the toxic side effects of existing drugs. However, recent developments in proteomics and bioinformatics will also allow predictions for drug MOAs as well as recommendations for better small molecule structures in the near future.

4.2 Expression Systems

If only for historical reasons *Escherichia coli* is still one of the best understood (micro) organism and this mainly because the first heterologous or recombinant protein expression was possible in this bacteria [30]. For the same reason, the molecular physiology and genetics of more interesting microorganisms for the industrial production of value products and protein expression, like *Bacillus*, *Corynebacterium*, *Aspergillus*, and yeast were not as advanced and therefore the mechanisms for the optimization of the expression yield in host organisms were not well understood at that time. Classical microbiology including strain collection and screening, random mutation and selection were the only way for identifying organisms suitable for production processes. At the beginning of recombinant biotechnology, it was considered an advantage that overexpression of recombinant (fusion) proteins in *E. coli* often lead to the formation of so called inclusion bodies, insoluble and microscopically visible protein particles containing the majority of the product yield. On the one hand, product formation as measured by SDS-PAGE could be correlated directly to the development of inclusion bodies in bacterial cells; on the other hand inclusion bodies could be separated by low speed centrifugation allowing the use of existing equipment for the initial enrichment step. However, several problems along the whole production process – from initiation of expression to the amount of highly purified end product – are caused by choosing *E. coli* and inclusion bodies as a strategy for protein expression. Besides the need for high concentrations of strong denaturing agents and chemicals, e.g., formic acid, urea, guanidine hydrochloride, SDS, which are needed for solubilizing inclusion bodies to make them amenable to all of the further purification steps, the high yield protein expression combined with intracellular formation of these particles obviously results in such a metabolic stress for *E. coli* that a lot of unwanted modifications occur and charge as well as size heterogeneity of the recombinant protein product can be detected. This in return heavily influences product yield as a whole and the ratio between possible yield (calculated on the expression band) and the final yield of the purified end product [31, 32].

4.3 Biotechnological Processes and 2-DE Applications

One of the first who applied 2-DE to the study of *E. coli* proteins was O'Farrell himself. He already worried about possible artifacts and observed charge heterogeneity of some of the protein spots [13]. However, some of the heterogeneity could have been introduced by his experimental strategy by using radioactive labeling of proteins and autoradiography for quantification. Labeling was done by adding a mix of not further described labeled ^{14}C-amino acids to the media over a not precisely defined time period. In general, the focus of his paper is mainly on the introduction of his new technology and some of the down-stream problems like quantification of autoradiograms by microdensitometry. Very important parameters, however, were neglected or not mentioned in the publication; e.g., precise culturing conditions, growth stages of bacterial cells with or without T4 infection were not recorded. Rinas [31] was the first who published on the product heterogeneity of proteins in recombinant *E. coli* and observed that overexpression of any recombinant protein – even if it is native to *E. coli* – can overburden the cellular machinery correctly to transcribe, translate and to fold the expression product into the correct structure, and therefore leads to product variants. In our own studies [32] we monitored, for the first time, fermentation kinetics and development of product yield by means of 2-DE. We were able to show a time dependent appearance of charge and size heterogeneity of insulin fusion proteins produced in *E. coli*. A monospecific polyclonal antibody against the last eight amino acids of the insulin A-chain, YQLENYCN, was used in Western-blot analysis of 2-DE separations of the crude fermentation product [33]. It could be shown that, at least for both of the insulin fusion proteins studied, no C-terminal degradation occurs during fermentation. This was a very important finding since the *N*-terminal fusion part – in this case either a part of β-galactosidase or interleukin 2 – has to be removed anyway, but C-terminal amino acid degradation would have a direct and severe impact on the product yield. Therefore, the individually separated, charge-heterogeneous, and Western-blot positive protein spots were analyzed by *N*-terminal sequencing after 2-DE and semi-dry-blotting on PVDF-membranes. *N*-terminal degradation up to 3 amino acids has been observed for both recombinant protein constructs, which, however, cannot account for the charge-heterogeneity seen. Taking into account that some oxidation reactions at several cysteine and methionine residues can also occur, at least most of the detected charge-differences should have no influence on product yield. Size heterogeneity, however, coming from the formation of dimers and trimers of the fusion construct obviously results in a decrease of product yield, since those oligomers are covalently linked – probably by the formation of Lys-Glu cross-links – and, dependent on the fermentation conditions, can represent up to one-third of the immunodetectable product which cannot be recycled. Finally, a finding, which was communicated by Hancock at Anabiotec '90 held in San Francisco, was confirmed by amino acid analysis of the insulin fusion protein spots. He and his colleagues at Genentech observed the introduction of norleucine in place of methionine into overexpressed recombinant proteins produced in *E. coli*. This phenomenon was contradictory to the purification strategies of the companies involved at that time, i.e., Eli Lilly, Genentech and Hoechst, which involved cyanogen bromide cleavage (or cyanogen chloride in the Hoechst process) at a methionine between the fusion part, e.g., β-galactosidase moiety, and the product part. The presence of a norleucine residue at this position protects the fusion protein from BrCN attack and leads to significant decrease in product yield. All the above effects and their significant impact on product quality and yield are more or less due to large scale fermentation processes and are normally not detected in recombinant bacteria grown in shaker flasks.

4.4 Proteomics and the Physiology of Microorganisms

Most of the studies published so far have put their main focus on the improvement of the down stream technologies for proteome analysis. Research on the up-stream side, however, was left for a long time to just a few academic groups. In particular, the work of Neidhardt [34], later together with van Bogelen [35] [1], Hecker [36] [2], Boucherie [37] and some other groups [38–40] was focused on the relevance of cultivation parameters on the cell physiology of microorganisms, i.e., cell growth, synthesis, and fluxes of primary as well as secondary metabolites, regulatory networks, and stress response. They all immediately realized the huge potential of 2-DE for the study of cellular physiology, by simply comparing and subtracting protein patterns which represent a certain metabolic state – precisely defined by careful recording of the experimental parameters and results – followed by correlating the patterns with the descriptive parameters and extracting the relevant information for (i) improvement of process efficiency and produc-

[1] Contributing author of this volume.
[2] Co-editor and contributing author of this volume.

tivity, (ii) characterization of regulatory protein networks, and (iii) identification of molecular targets for drugs and metabolic design. What was proposed by the Andersons [18, 41–47] in the early 1980s for man and which was not possible until now – after completion of the Human Genome Project – Neidhardt and his colleagues started to work out this information for bacteria in 1978, long before the *E. coli* genome was sequenced.

4.5
Streptomyces and Cycloheximide Biosynthesis

Besides the common problems for the application of 2-DE, e.g., low reproducibility, laborious and expensive, lack of skilled personnel, Dykstra and Wang [48] used 2-DE to study the intracellular protein profile of *Streptomyces griseus* in relation to cycloheximide biosynthesis. Four proteins were found to be specifically repressed by the antibiotic and product yield could be increased twofold by simple addition of a neutral resin to the culture broth and adsorption of cycloheximide.

4.6
Proteome Analysis for the Identification of Key Enzymes for Metabolic Engineering

4.6.1
Ethanol Fermentation

Zymomonas mobilis is an anaerobic, Gram-negative bacterium, which is widely used in tropical regions for the fermentation of alcoholic beverages. Unlike most other plants and fungi it utilizes the Entner-Doudoroff pathway for the conversion of glucose to pyruvate. The organism is of commercial interest due to its highly active pyruvate decarboxylase, which is used in biotransformation processes for the production of fine chemicals. Ingram et al. investigated the glycolytic and fermentative pathways in *Z. mobilis* with 2-DE with a special focus on the ethanologenic enzymes. Alcohol dehydrogenase II was identified as a prominent stress protein and three other acidic stress proteins could be compared in size and gel position with *E. coli* DanK, GroEl, and GroEs [49]. With the knowledge generated by this and other studies, it will not only be possible to increase ethanol production by metabolic engineering and metabolic design, but also to identify new enzymes for biocatalysis and to construct recombinant microorganisms for commercially interesting biotransformation processes.

4.6.2
Biotin Biosynthesis

In their recent study, Shaw et al. [50] successfully applied 2-DE to detect enzymes of the biotin biosynthetic pathway from wild type *E. coli*. They identified the biotin synthase step as the major flux control point and as an important key step for any commercial fermentation process. By the introduction of the *E. coli bio* operon via a broad-host range plasmid they transformed *Agrobacterium/Rhizo-*

bium HK4 which then produced 110 mg/l of biotin in a 2-l fermentor. This can be considered a major achievement towards a commercially viable biotechnological production process for biotin, which should be in the range of 1 g l^{-1} day^{-1}. Presently this commercially very interesting vitamin, with pharmaceutical, nutritional and cosmetic application and a market size of more than $100 million per year, is synthesized on an industrial scale by multi-step chemical processes. However, biotechnological production could be advantageous in terms of cost, simplicity, and environmental considerations.

4.6.3 *Antibiotic Production*

Fosfomycin [(–)-*cis*-1,2-epoxypropylphosphonic acid], first discovered in *Streptomyces fradia*, inhibits the initial reaction in biosynthesis of prokaryotic peptidoglycans and thus has broad-spectrum antibiotic activity against Gram-positive and Gram-negative bacteria. Watanabe et al. [51], while working on the improvement on the biotransformation process of *cis*-propenylphosphonic acid (cPA) to fosfomycin with *Penicillium decumbens*, used 2-DE to monitor all the proteins which are induced by cPA. A 31-kDa protein (EpoA) was both cPA induced and overaccumulated in a strain which more efficiently converted cPA. After purification and cloning of the EpoA gene (epoA) and subsequent subcloning into *P. decumbens*, a fourfold increase in epoxidation activity and product formation was achieved. epoA disruption mutants, however, could not transform cPA into fosfomycin.

4.7 Characterization of Mammalian Cell Cultures Used in Fermentation

The fermentation of mammalian cell cultures is of high importance for the production of a huge variety of monoclonal antibodies for diagnostic and therapeutic use as well as other important protein drugs like tissue plasminogen activator (tPA). Two examples of long-term in vitro culture – a repeatedly subcloned non recombinant human melanoma cell line expressing natively human tPA and recombinant tPA producing Chinese hamster ovary (CHO) cells – have been investigated with 2-DE by Harant et al. [52]. Protein pattern comparison of the secreted cellular proteins was used to monitor the changes which occur over a period of 14 months of continuous fermentation and those patterns were than compared to the mother cell line recultivated after being frozen for more than two years. In the attempt to monitor the physiological consistency of animal cells in culture by 2-DE it could be shown for the first time that the comparison of secretory protein patterns from nonrecombinant and recombinant reflect a qualitative and quantitative picture of the physiological state of cell lines.

The productivity, the proteome, and protein phosphorylation in response to lower temperature was investigated in CHO cells engineered to synthesize the model product secreted alkaline phosphatase [53]. According to flow cytometric analysis (FACS) 80% of the cells accumulated to the G1 phase after a temperature shift from 37 to 30 °C. The G1/S transition is the most important restriction point in the mammalian cell cycle and its control is essential in many cellular processes such as embryonic development and cancer. As expected, temperature reduction introduces changes in the overall protein pattern and tyrosine phosphorylation. More importantly, low temperature cultivations lead to a 3.5-fold higher product titer of secreted alkaline phosphatase with respect to fermentation at 37 °C. A decrease in cultivation temperature seems to suppress cell growth obviously by halting the cell cycle in G1, which may result in a delayed onset of apoptosis.

Due to the high commercial interest in cultivated CHO cells, it will be necessary to characterize all the responsible cellular mechanisms which result in a higher productivity on the one hand, and changes in growth rate as well as nutrient, e.g., glucose, consumption on the other.

The creation of 2-DE reference maps of protein patterns (***normalome***) with mammalian cells is dependent on the possibilities for the cultivation in serum free media. These reference maps are very useful for further investigations of physiological responses to modifications in culture conditions, cell line improvement, and culture supernatant protein content. In particular, the intracellular processes which are induced or repressed by the expression of recombinant proteins in CHO cells and which are reflected in a characteristic proteome have to be identified. Champion et al. subjected CHO cells [54], harvested from a 2-l fermentor and during the exponential growth phase, to proteome analysis. Relevant cytosolic proteins, such as HSC70 and peptidyl *cis-trans* isomerase, both of which play a role in cellular stress response and proper protein folding, as well as polypeptides from the endoplasmatic reticulum (ER), like GRP78, protein disulfide isomerase, calnexin and calreticulin, and mitochondrial marker proteins, e.g., HSP60, were some of the 25 protein spots unequivocally identified in this study and can serve as useful landmarks for protein pattern comparisons.

Cell-free protein translation has been a routine synthesis technique in molecular biology laboratories for several decades. Most advantageous applications include the generation of toxic proteins and the introduction of novel or derivatized amino acids into newly synthesized polypeptides. 2-DE was used in a recent study by Schindler et al. [55] to investigate the protein composition of an *E. coli* cell extract employed in in vitro translation experiments. Green fluorescent protein (GFP) was chosen as a target protein and in vitro synthesis was monitored by a standard fluorescent assay.

4.8 Dairy Products and Beer Brewing

The dairy and the beer brewing industries [56–61] rely on reproducible, controlled and predictable fermentations, but variations in the starter culture performance are not uncommon and require reliable and fast methodology for process monitoring. In addition, little is known about the physiology of the early lag phase, whereas process economics would significantly improve by a reduction of the lag phase, making a faster and better-controlled fermentation feasible. In order to learn more about "early" protein synthesis, this phenomenon was studied in *Lactobacillus delbrueckii* ssp. bulgaricus which in combination with *Streptococcus thermophilus* is mainly used for yogurt production as well as a starter culture for Swiss-type cheese. The proteomes of exponentially growing and stationary cells in their natural environment, i. e., milk, were compared and several characteristic proteins were identified and sequenced [56]. An approach to elaborate a reference map of soluble proteins of *Streptococcus thermophilus* has been started by Perrin et al. [57] and underlines the importance of such studies. Proteolysis during cheese making was studied with 2-DE by Chin and Rosenberg [59] aimed at the better understanding of proteolysis-related variables and isolation of individual polypeptides relevant for ripening and quality.

The brewing industry is a field where traditional and new technologies coexist successfully. Recent studies have shown [60, 61] that proteome analysis of brewer yeast provides valuable information for process improvement, e.g., fed-batch cultivation with a high aeration flow rate.

As stated by Kuipers [58], food biotechnology in general will benefit greatly from proteomics and functional genomics approaches. The new technologies will create novel opportunities to ensure the safety of foods, to improve quality and economics of fermentation products at the same time, and to substantiate health claims related to the ingestion of specific microbes as well as their respective influence on existing microflora. Several interesting investigations on specialty foods, fruits, and wine are underlining this statement.

4.9
Plant Biology and Productivity of Culture Plants

The Andersons recommended 2-DE for the study of wheat seed proteins as early as 1985 and to use the resulting protein patterns for variety control and patent protection. Over 900 papers applying 2-DE methodology to a broad variety of plants and phytobiological problems have been published since then. Compared to other biomaterials, 2-DE studies of plants and plant-derived material in general are more complex. The rigid cell wall components, the relatively low overall protein content, and the accompanying polyphenols make special sample pretreatment protocols necessary. Thanks to the recent advances in the techniques for identifying proteins separated by 2-DE and in methods for large-scale analysis of proteome variations, proteomics is becoming an essential methodology in various fields of plant biology. In the study of pleiotropic effects of mutants and in the analysis of responses to hormones and to environmental changes, the identification of involved metabolic pathways can be deduced from the function of affected proteins [67]. In molecular genetics, proteomics can be used to map translated genes and loci controlling their expression, which can be used to identify proteins accounting for the variation of complex phenotypic traits. Linking gene expression to cell metabolism on the one hand and to genetic maps on the other, proteomics has grown rapidly to become a key technology in plant research.

Since the development of a full range cacao aroma and taste precursors is likely to occur during fermentation – unfermented beans do not provide the characteristic aroma upon roasting – Lerceteau et al. [62] studied the fermentation process of cacao beans (*Theobroma cacao*) with 2-DE prior to roasting. As proteolysis seems to be essential for the formation of cacao flavor, they focused their work on the first seven days of fermentation and the identification of lower molecular weight polypeptides with relevance to cacao flavor. Experimental 2-DE approaches were also applied to learn about gene expression in the mango fruit ripening process [63] as well as the analysis of muscadine wine proteins and their possible influence on wine stability and clarity [64, 65].

4.10 Fish

Another application of 2-DE in the area of food and food processing was published by Morzel et al. [66]. Changes in proteins influence to a very large extent the quality of fresh or processed fish products, particularly texture attributes. In this current study, salmon fillets treated with *Lactobacillus sake* were compared with untreated material. Endogenous enzymes are responsible for most quantitative changes, however, fermentation has a significant effect only on proteins with pI 6.25–8.25. Tropomyosin was found to be a suitable substrate for *L. sake*.

4.11 Symbiosis

Plant microbe interaction, e.g., legume-rhizobium symbiosis, is important for world agriculture [68, 69] since biological nitrogen fixation via the legume-rhizobium symbiosis is cost-effective, and avoids the often uneconomic use of nitrogenous fertilizers. However, low pH in soil decreases the productivity of legume crops and pastures due to the adverse effects of acidic soil on rhizobium. Therefore, acid-tolerant strains of root nodule bacteria from the Mediterranean rim were identified and are now successfully used in Western Australia on large areas of acidic soils. Mutagenesis and proteome analysis was employed to identify the up to 50 genes essential for rhizobium growth at low pH.

Besides bacteria which are beneficial for plant growth and development, several plant pathogenic bacteria are known which are the cause of sometimes devastating losses in fruit production. In combination with classical microbiological methods, PCR and proteome analysis, a new Erwinia sp., *Erwinia pyrifoliae*, was characterized [70].

5 Technology Developments, Market Potential, and Outlook

When the first genome project – the yeast genome – was completed it became apparent that proteomics would develop into one of the most important platform technologies in life sciences. In addition, despite the speed of the technological and application developments, especially in mass spectrometry and automation [71–74], proteomics is still in its infancy and most publications in the proteomics field are still on the exploratory side. However, some groups have put increased emphasis on product and process oriented use of the technology.

The concept "From Genome to Proteome" of the pace making biennial Siena meetings – the first one was held in 1994 – is now generally accepted as a guideline in life science research. We envisage exponential growth in all scientific fields needed for the advancement of proteome research. This trend is reflected not only by the number of publications and patents, but also by an increase in governmental funding of academic proteome projects as well as founding of start-up companies.

Until 1998, worldwide, there was just a small group of scientists, mainly in academia, who had realized the importance and the potential of proteomics. One could hardly count more than ten companies worldwide.

Just recently, especially pharmaceutical companies, desperately looking for new and unique specific pharmacological targets, started to realize the huge potential of proteomics. As expected, this led to an increase in financial investment and therefore to a tremendous push in development as well as acceptance of this biotechnological area [19, 20, 75]. Large pharmaceutical companies are now building up their own individual activities in proteomics. In addition, alliances with proteomics companies are formed or existing collaborations with those genomics companies which have actively adopted proteomics research programs are intensified. Table 1 gives a representative overview of the business dynamics of the last two years.

Different older and very young companies are now exploiting the interest of the pharmaceutical industry to investigate protein abundance and interaction in the context of disease. In addition, some of the former equipment manufacturers, e.g., Amersham Pharmacia Biotech, changed their business strategies towards the elaboration and marketing of fully integrated proteomics production lines, or they placed a direct investment into promising start-up companies, e.g., Bruker's engagement in GeneProt. It is obvious today that the first company providing a fully automated device for quantitative proteomics – from sample preparation to problem solution – will have an impact on future development in biotechnology comparable to Genentech, Amgen, Millennium, and Celera.

However, provided that a fully automated proteome analysis machine with a throughput like the sequencing devices used by Celera for the human genome project will be available soon, we will be by no means able handle, to store, or to retrieve the information overflow. It is obvious already that existing concepts in bioinformatics and available software tools were not developed for and are therefore not suited for the handling of proteome data. Nature functions by integration, and the adoption of a more holistic view of complex biological systems is key to the development of better bioinformatics. To get the most from proteome data, we need to take account of information on the regulation of gene expression, metabolic pathways and fluxes, transport and turnover rates, signaling cascades, etc. Proteins do not work in isolation – approximately 10,000 different proteins in varying concentrations have to work together in every mammalian cell – but are involved in interrelated networks. In particular, the understanding of the linguistics – the language, grammar, and spelling of proteins communications – of protein-protein-interaction, and also all the other intracellular interactions of biomolecules, e.g., protein-DNA, protein-RNA, etc., will be vital to our understanding of normal and abnormal cell development [76], and will allow us to create an integrated mapping between genotype and phenotype.

Table 1. Representative overview of the business dynamics of the last two years

Company	Business activities	Corporate Partners
Amersham Pharmacia, Biotech, USA	Supplier of genomic and proteomic systems, not active in drug development	Zeneca, Dyax Corp., Procter & Gamble
Applera (Merger of two divisions of Applied Biosystems and Celera), USA	Proteomic research tools, molecular biology, bioinformatics and biomolecule detection	Oxford Glycosciences, Geneva Proteomics, Millipore Corporation
Curagen	Yeast 2 hybrid	
Geneva Proteomics	Bioinformatics and data warehouses	Novartis, Compugen
(GeneProt)		Bruker, Daltonics
Genomic Solutions	GeneTAC biochip system, gene expression analysis Investigator proteomic system	Affymetrix, Introgen
GPC Biotech	One stop shop for genomic and proteomic solutions	Evotec Biosystems, Morphosys, Aventis, Atugen, Bayer, Boehringer Ingelheim, Byk Gulden
Hybrigenics, Paris, France	High throughput yeast two hybrid	Lynx Therapeutics
Large Scale Biology	ProGEx, high-throughput high-resolution proteomics platform, geneware, gene expression in plants	Dow Chemical, Gemini Genomics
MDS Proteomics (formed of Protana, Ocata and MDSMCD Inc)	Study of protein-protein interactions, drug development	NK
Myriad Genetics, USA	ProNet, positional cloning and protein interaction program, yeast two-hybrid	Bayer, Schering-Plough
Oxford Glycosciences	ProteoGraph, high-throughput proteomics	Pfizer, Merck, Monsanto, GlaxoSmithKline, Upjohn, Bayer, Medarex
Proteome Sciences, UK	Proteomics technologies for diagnostics and drug development	Aventis
Proteome Systems Ltd., Australia	Automated proteomics machine	Sigma-Aldrich, Shimadzu, Millipore Corp.
Rigel	Yeast 2 hybrid	NK
Xzillion (former Aventis Research & Technologies GmbH)	Proteomics technology platform	Brax, UK

Proteomics, and to a much higher extent genomics-based drug discovery, is dependent on precise and unequivocally functional annotation. Bioinformatics has to deliver highly integrated, interoperable, and flexible databases – "data warehouses." Since an infinite number of proteomes can be – theoretically – obtained by any genome, a fully integrated biological data base or data warehouse has to allow the user to store and to retrieve an infinite number of descriptive parameters[3] (Dp). Some of those Dps will be hard data, e.g., DNA sequences, and some will be derived data or meta-data, e.g., protease stability or cellular localization. Present data base architecture has severe limitations with respect to demands of proteomics research. On the other hand, any data to be stored has to meet the highest possible quality standards, which will create major problems for the experimentalist with respect to accurate recording of all Dps. The more genome, proteome, and transcriptome annotation is automated, the greater will be the need for tight collaboration between software developers, annotators, and experimentalists.

Presently, companies like BASF, Bayer, Degussa, Novartis, and others involved in larger scale biotechnological production processes, e.g., vitamins, amino acids, organic acids, antibiotics, have started in-house proteomics projects to optimize their respective fermentation processes. Due to the lack of fast, sophisticated, and reliable methods for studying cell physiology and cellular regulatory networks of organisms in fermentation processes, this field was neglected for a long time. Furthermore, most attempts to increase overall product yield by overexpression of selected proteins of the respective biosynthetic pathway failed or gave no advantage over classical mutation-selection approaches. Now, having access to bacterial and yeast genome sequences, DNA chip technology, mass spectrometry, and 2-DE on the one hand, and on the other hand the knowledge about physiological regulatory networks in response to stress and nutrients, much more straightforward strategies can be designed to increase product yield and purity as well as the overall economics of fermentation processes.

The market size for products from fermentation processes is in the range of US$ 25–30 in sales:

- Pharma proteins, e.g., EPO, insulin, human growth hormone, β-Interferon, G-CSF
- Antibiotics
- Steroids, lipid lowering drug precursor, animal health antibiotics, vitamins
- Organic acids
- Technical enzymes
- Amino acids

where antibiotics account for more than 60% of the sales, but pharma protein deliver the best profits and the highest growth dynamics.

[3] Descriptive parameters = any experimental data describing a biological system, e.g., species, age, sex, temperature, cell count, protein concentration, isoelectric point, molecular weight, protein sequence, protein structure, enzyme kinetics, etc.

6 Conclusion

Proteome studies have now become feasible and – with vast speed – proteomics has entered the biotech business arena. It is presently one of the fastest growing branches in biotechnology. However, as always with new exiting technological developments, you see a lot of hype and, therefore, utmost caution is advised when investing in in-house or external proteomics projects. Some of the companies mentioned above offer such excellent state-of-the-art services in proteomics and protein analysis that the build-up of in-house capacities should be carefully evaluated. In most cases, principal scientists in those companies have spent decades in their field and show a proven track record as proteomics experts. In addition, as in the bioinformatics field, dedicated experts in protein chemistry and mass spectrometry are very rare.

It should be clearly pointed out here that proteomics is not a product as such. It is a highly sophisticated technological area that can be used to improve processes and product safety, and to identify new products. First, proteome analysis produces raw data that are transformed into information by linking raw data with the descriptive parameters (Dps) of the respective experiment. Second, this information will be turned into knowledge by the intelligent and straightforward application of this information on the respective biological problem. Third, this knowledge will give a competitive advantage and will generate value by that means.

7 References

1. Letsinger RL, Mahadevan V (1965) J Am Chem Soc 87:3526
2. Letsinger RL, Caruthers MH, Jerina DM (1967) Biochemistry 6:1379
3. Merrifield RB, Stewart JM (1965) Nature 207:522
4. Merrifield RB (1965) Science 150:178
5. Morrow JF, Berg P (1972) Proc Natl Acad Sci USA 69:3365
6. Kelly TJ, Smith HO (1970) J Mol Biol 51:393
7. Smith HO, Wilcox KW (1970) J Mol Biol 51:379
8. Meselson M, Yuan R (1968) Nature 217:1110
9. Szostak JW (1993) Nature 361:119
10. Eigen M, Gardiner W (1984) Pure Appl Chem 56:967
11. Szostak JW, Bartel DP, Luisi PL (2001) Nature 409:387
12. Klose J (1975) Humangenetik 26:231
13. O'Farrell PH (1975) J Biol Chem 250:4007
14. Bjellqvist B, Ek K, Righetti PG, Gianazza E, Gorg A, Westermeier R, Postel W (1982) J Biochem Biophys Methods 6:317
15. Aebersold RH, Teplow DB, Hood LE, Kent SB (1986) J Biol Chem 261:4229
16. Eckerskorn C, Lottspeich F (1989) Chromatographia 28:92
17. Karas M, Hillenkamp F (1988) Anal Chem 60:2299
18. Anderson NL, Anderson NG (1979) Behring Inst Mitt 63:169
19. Müllner S, Neumann T, Lottspeich F (1998) Arzneim-Forsch/Drug Res 48:93
20. Lottspeich F (1999) Angew Chem Int Ed 38:2476
21. Anderson NL, Esquer-Blasco R, Anderson NG (1994) Methods in toxicology. In: Tyson CA, Frazier JM (eds) In vitro toxicity indicators. Academic Press, p 463

22. Aicher L, Wahl D, Arce A, Grenet O, Steiner S (1998) Electrophoresis 19:1998
23. Grenet O, Varela MC, Staedtler F, Steiner S (1998) Biochem Pharmacol 55:1131
24. Anderson NL, Esquer-Blasco R, Richardson F, Foxworthy P, Eacho P (1996) Toxicol Appl Pharmacol 137:75
25. Anderson NL, Swanson M, Giere FA, Tollaksen SL, Gemmell A, Nance SL, Anderson NG (1986) Electrophoresis 7:44
26. Steiner S, Wahl D, Mangold B, Robison R, Raymackers J, Meheus L, Anderson L, Cordier A (1996) Biochem Biophys Res Comm 218:777
27. Anderson NG, Anderson NL (1998) Electrophoresis 19:1853
28. Dax CI, Lottspeich F, Müllner S (1998) Electrophoresis 19:841
29. Mangold U, Dax CI, Saar K, Schwab W, Kirschbaum B, Müllner S (1999) Eur J Biochem 266:1184
30. Goeddel DV, Heyneker HL, Hozumi T, Arentzen R, Itakura K, Yansura DG, Ross MJ, Miozzari G, Crea R, Seeburg PH (1979) Nature 281:544
31. Rinas U (1992) DECHEMA Biotechnol Conf 5:529
32. Müllner S, Karbe-Thönges B, Tripier D (1993) Anal Biochem 210:366
33. Müllner S, König W, Neubauer HP (1991) J Immunol Methods 140:211
34. Pedersen S, Bloch PL, Reeh S, Neidhardt FC (1978) Cell 14:179
35. Neidhardt FC, Wirth R, Smith MW, van Bogelen R (1980) J Bacteriol 143:535
36. Hecker M, Wachlin G, Dunger AM, Mach F (1984) FEMS Microbiol Lett 25:57
37. Bataille N, Thoraval D, Boucherie H (1988) Electrophoresis 9:774
38. Schweder T, Kruger E, Xu B, Jurgen B, Blomsten G, Enfors SO, Hecker M (1999) Biotech Bioeng 65:151
39. Tobisch S, Zuhlke D, Bernhardt J, Stulke J, Hecker M (1999) J Bacteriol 181:6996
40. Blankenhorn D, Phillips J, Slonczewski JL (1999) J Bacteriol 181:2209
41. Anderson NL, Anderson NG (1977) Proc Nat Acad Sci USA 74:5421
42. Anderson NG, Anderson NL (1982) Clin Chem 28:739
43. Anderson NG, Anderson NL (1982) Med Lab 11:75
44. Anderson NG, Anderson NL (1981) In: Keenberg M (ed) Proceedings of the 1981 Battelle Conference on Genetic Engineering Reston VA, p 163
45. Anderson NL (1982) Trends Anal Chem 1:131
46. Anderson NG, Anderson NL (1985) Am Biotech Lab Sept/Oct:4
47. Anderson NL (1985) In: Stevenson RE (ed) Uses and standardization of vertebrate cell cultures in vitro. Monograph No 5, Tissue Culture Association, p 189
48. Dykstra KH, Wang HY (1990) Appl Microbiol Biotech 34:191
49. An H, Scopes RK, Rodriguez M, Keshav KF, Ingram LO (1991) J Bacteriol 173:5975
50. Shaw NM, Lehner B, Fuhrmann M, Kulla HG, Brass JM, Birch OM, Tinschert A, Venetz D, Venetz V, Sanchez JC, Tonella L, Hochstrasser DF (1999) J Ind Microbiol Biotechnol 22:590
51. Watanabe M, Sumida N, Murakami S, Anzai H, Thompson CJ, Tateno Y, Murakami T (1999) Appl Environ Microbiol 65:1036
52. Harant H, Wimmer K, Wenisch E, Strutzenberger K, Reiter M, Bluml G, Gaida T, Schmatz C, Katinger H (1992) Cytotechnol 8:119
53. Kaufmann H, Mazur X, Fussenegger M, Bailey JE (1999) Biotechnol Bioeng 63:573
54. Champion KM, Arnott D, Henzel WJ, Hermes S, Weikert S, Stults J, Vanderlaan M, Krummen L (1999) Electrophoresis 20:994
55. Schindler PT, Macherhammer F, Arnold S, Reuss M, Siemann M (1999) Electrophoresis 20:806
56. Rechinger KB, Siegumfeldt H, Svendsen I, Jakobsen M (2000) Electrophoresis 21:2660
57. Perrin C, Gonzalez-Marquez H, Gaillard JL, Bracquart P, Guimont C (2000) Electrophoresis 21:949
58. Kuipers OP (1999) Curr Opin Biotechnol 10:511
59. Chin HW, Rosenberg MJ (1998) Food Sci 63:423
60. Gorinstein S, Zemser M, Vargas-Albores F, Ochoa JL, Paredes-Lopez O, Scheler C, Salnikow J, Martin-Belloso O, Trakhtenberg S (1999) Food Chem 67:71
61. Joubert R, Brignon P, Proth J, Boucherie H, Gendre F (2000) Monogr Eur Brew Conv 28:171

62. Lerceteau E, Rogers J, Petiard V, Crouzillat D (1999) J Sci Food Agric 79:619
63. Chaimanee P, Suntornwat O, Lerrtwikool N, Bungaruang L (1999) J Biochem Mol Biol Biophys 3:75
64. Lamikanra O, Inyang ID (1987) Fla State Hortic Soc 99:148
65. Lamikanra O, Inyang ID (1988) Am J Enol Vitic 39:113
66. Morzel M, Verrez-Bagnis V, Arendt EK, Fleurence J (2000) J Agric Food Chem 48:239
67. Kuwabara C, Arakawa K, Yoshida, S (1999) Plant Cell Physiol 40:184
68. Glenn AR, Reeve WG, Tiwari RP, Dilworth MJ (1999) Novartis Found Symp 221:112
69. Worland S, Guerreiro N, Yip L, Djordjevic MA, Djordjevic SP, Weinman JJ, Rolfe BG (1999) Aust J Plant Physiol 26:511
70. Rhim SL, Volksch B, Gardan L, Paulin JP, Langlotz C, Kim WS, Geider K (1999) Plant Pathol 48:514
71. Washburn MP, Wolters D, Yates JR (2001) Nat Biotechnol 19:242
72. Yates JR (1998) J Mass Spectrom 33:1
73. Goodlett DR, Bruce JE, Anderson GA, Rist B, Pasa-Tolic L, Fiehn O, Smith RD, Aebersold R (2000) Anal Chem 72:1112
74. Jankowski J, Stephan N, Knobloch M, Fischer S, Schmaltz D, Zidek W, Schlüter H (2001) Anal Biochem 290:324
75. Anderson NG, Anderson NL (1996) Electrophoresis 17:443
76. Attwood TK (2000) Science 290:471

Received: April 2002

Adv Biochem Engin/Biotechnol (2003) 83: 27–55
DOI 10.1007/b11112

Probing the Molecular Physiology of the Microbial Organism, *Escherichia coli* Using Proteomics

Ruth A. VanBogelen

Pfizer Global Research and Development, 2800 Plymouth Rd, Ann Arbor, MI 48105, USA
E-mail: Ruth.Vanbogelen@pfizer.com

Genomics and proteomics technologies have yielded volumes of data for more than 20 years, and they continues to produce data at an astounding rate. Has all of this data helped us understand more about life, or it is just bogging us down in details that cannot be assembled into meaningful ideas? This review of the proteomics efforts over the last couple of decades is meant to emphasize that a new scientific discipline has emerged, Molecular Physiology, and that, indeed, this discipline is contributing to our understanding of life. Molecular physiology offers the reductionisms details of individual cellular molecules and offers the systems biology multivariant and high-dimensional datasets of cellular molecules.

Keywords. Proteomics, Physiology, Microbiology, *Escherichia coli*, Two-dimensional electrophoresis, Stress responses

1 Introduction

The mystery of how biological systems function has been the impetus for much research for over 20 years. Biologists have addressed this problem in a reductionist approach by studying in detail many cellular molecules. Engineers have approached this problem by viewing the biological system as a closed subsystem approachable with mathematical descriptors. A few physiologists have tried to describe the biological systems by studying stimulus-response behavior of cells using proteomics techniques [1].

Much knowledge about "the hows" of proteomics has been gained by many trial-and-error experiments. Standardized conditions for designing experiments and for running gels have been implemented for *E. coli* [2]. Much experience has been gained from analyzing the image data. The attempts to collate all of this data together collectively into a comprehensive database have revealed a logical organization scheme for grouping the experimental data into different projects and maps so that logical queries of the data can be done [3]. In addition to the methods for doing the work and analyzing the results, there was also a need to build a vocabulary with definitions of words useful in the description of the work summarized in [4].

Most importantly, much has also been learned about what global proteomic monitoring can and cannot reveal about the cell physiology. Intriguing discoveries have emerged from the experiences using the global analysis of protein expression profiles. Two discoveries may greatly contribute to the understanding of biological systems [4]. First, the physiological behavior of cells can be diagnosed by changes in protein expression. Second, although the potential matrix of behaviors for biological systems would appear to be enormously large (even for bacterial cells with less than five thousand genes), there is growing evidence that biological systems might be described by a relatively small number of "physiological modules" of behaviors.

This chapter will describe how proteomics can help with the elucidation of the molecular physiology of the microbial organism, *Escherichia coli*. The reader might want to skip to the Concluding Remarks for a vision of where microbial physiology should be in ten years and then return to the first seven sections which discuss where we are now and what is needed to understand microbial physiology at a molecular level. The first section discusses the current technology limitations of proteomics. The next four sections discuss methodologies for proteome studies and present examples of these studies. The sixth section begins the discussion of how the molecular physiology data generated with 2D gel studies can be used for diagnosing the physiological states of cells. The seventh section briefly introduces the idea of switching from wet lab experiments to dry lab experiments using mathematical structures of biological data in the form of simulations and cell models.

2 Proteomic Technologies Were Introduced Twenty-Five Years Ago, but Still Have Many Limitations that Hinder Studies of Complex Mixtures of Proteins in Cells

In 1975, two-dimensional polyacrylamide gel electrophoresis (2D gel) was introduced as a methodology for separating complex mixtures of cellular proteins [5]. Almost immediately the method captured the attention of a few scientists around the world who viewed it as a tool for solving the mystery of what happens to cellular protein expression in the ever-changing environment of an organism. Despite the flurry of papers that reported exciting and new information about individual cellular proteins in bacteria, yeast, plants, and animals, both the methodology and the information yielded were not widely appreciated until the 1990s [6]. The word proteome was introduced by Wilkins [7] and coworkers and was defined as the "protein complement of the genome". Although this word is a static description of the cell's complement of proteins, the word proteomics [4, 6] connotes the studies of the dynamic phenotypes of individual proteins. Two recent reports have presented arguments for why proteomic studies are needed despite the intensive efforts underway to sequence entire genomes and analyze global transcription changes [8, 9].

2D gels have been the method of choice for many years and much new biology has been discovered with the method. Despite the success of this method the limitations are numerous (see below) and much effort is being focused on developing a more robust method for global analysis of proteins.

2.1 Reproducibility of 2D Gels

When 2D gel methods were first introduced both the equipment and reagents (used to produce gels) were prepared by each laboratory. Several formats were introduced in the first five years [10]. This led to much variation in the quality of the protein separation and in the overall protein pattern. In the last ten years, reproducing 2D gels from lab to lab has been simplified because much of the equipment and many of the reagents needed for 2D gels are commercially available (Genomic Solutions; BioRad; Amersham). 2D gels are now a "kit" technology or can be done with biotechnology companies focusing on proteomics (e.g., Large Scale Proteomics).

2.2
Protein Identification

In the 1980s analytical methods for identification of proteins separated by 2D gels were introduced. First, micro-scale Edman degradation methods were described [11]. Later mass spectrometry methods were developed to identify proteins from 2D gels [12]. These analytical methods are now widely used [13, 14]. A large amount (greater than 100 femtomole) of protein is required to link the polypeptide to a gene on the chromosome. Many cellular proteins are not present at high enough concentrations to be analyzed by current technologies. The abundance of individual protein molecules in cells varies from less than 1 molecule per cell to over 10,000 molecules per cell, five to eight orders of magnitude [6]. However, regardless of the cell type, about 150 polypeptides account for 70% of the protein mass of the cell [15]. Analytical isoelectric focusing limits the amount of total protein that can be loaded, so typically a maximum of 300 proteins can be analyzed without some fractionation method to enrich for the less abundant polypeptides. A current approach is to fractionate the original cell material and use isoelectric focusing on narrow pH ranges [16] to enrich for each protein in the mixture. The information lost by this approach is the quantification of the each protein relative to the complex mixture. Although this is useful information, this is not the approach that will define the physiology of an organism.

2.3
Protein Solubility

Another technical problem in this area has been protein solubility. Proteins with several transmembrane domains have not been detected on 2D gels due to their poor solubility in the first dimension electrophoresis step. In studies to develop methods to increase the solubility of membrane proteins, investigators have found many other proteins that were not detected by the standard solubilization methods [17].

2.4
Protein with Extreme pIs and MW

A third problem has been the lack of standard methods to separate proteins with extreme isoelectric points and molecular weights. Gorg et al. have published methods to separate proteins with isoelectric points as high as 11.7 [18]. Detection of low molecular weight proteins can be done with mass spectrometry. Separation of high molecular weight proteins continues to be difficult. One recent advance that addresses this problem is the ICAT method being developed in Aebersold's lab [19]. In this method two protein samples are isotope tagged (one sample with the hydrogen form of the tag, a second sample with the deuterium form of the tag) on the cysteine residues of proteins. The proteins are subsequently cleaved with a protease (typically trypsin) and analyzed by mass spectrometry. The quantitation of peptides that differ in mass by 8 Da allows for an accurate determination of the ratio of that protein in the original samples. This method is likely to miss low molecular weight proteins that might not contain a cysteine peptide, but should give good quantification of very high molecular weight proteins that would be represented by several peptides.

2.5 Image Analysis Systems

A major technical difficulty has been with image analysis system [10]. Most of the early 2D gels studies done in the 1970s generated data (largely image data) on the dynamic properties of proteins, namely changes in the level or synthesis rates of individual proteins induced by different stimuli or conditions. These studies have provided many new findings about cellular proteins that relate the proteins to the workings of the cells rather than to the genome of the cell. However, few investigators have employed 2D gels for quantitative translation analysis studies.

Computer aided image analysis systems have been designed to convert the image data to numerical data and to allow for the data from multiple images to be merges by image matching strategies [20]. Several software systems are commercially available, but these software systems tend to be difficult to learn and are very labor intensive. Biotechnology companies have recognized the need for improved image analysis and several have intensive efforts underway to eliminate this bottleneck step. One system being developed reverses the order of the analysis. This system, called Z3 (Compugen, Inc.) registers images prior to spot detection [21]. The advantage of this system is the reduced time for manual editing. With current systems, the matching of two images takes hours; Z3 does this in minutes. Another approach is image searching (Scimagix, Inc.) that is very important for across experiment analysis. The unique feature of this system is that it will perform a search of features within an image across many 2D gel images and that it allows a matrix of gels and spots on gels that remains linked to the images.

2.6 Protein Detection

Radiolabeling of proteins has provided a means for excellent quantitative analysis of proteins over three and a half orders of magnitude. However for many microbial organisms, defined media (that allows radiolabeled amino acids to be incorporated into protein) have not been developed. For work with humans and many animals, the use of radioisotopes is not an option. This problem is being addressed by developing methods for fluorescent dying and staining of proteins [22, 23].

2.7 Data Mining

Tools and methods to explore quantitative variation of thousands of cellular proteins has not been solved [24]. Several reports have been published trying a variety of statistic approaches [25–27]. Just as with RNA expression data, there is much work to be done in this area.

3
Establishing Standardized Methods for Experiments are Important for Obtaining Meaningful Results from Proteomic Studies

When the goal of a proteomics project is to learn about the physiology of an organism, much thought and consideration must be given toward establishing standardized and systematic methodologies. This is true for studies of microbial organisms and multi-cellular organisms. Many investigators have been critical of 2D gels because the gels themselves are not considered reproducible (i.e., one gel can almost never be perfectly superimposed on a second gel). Nevertheless, it turns out this problem is not the most limiting factor in proteomic studies. Design and execution of experiments is most critical to success in this area just as it is in many area of biological research [28, 29].

One of the early standardization steps in the physiological studies of *E. coli* was the development of a defined medium that allows any nutrient to be depleted or replaced [30]. This medium, glucose minimal MOPS, has become the standard media for proteomic studies of *E. coli*. Supplements to the standard medium, including amino acids, vitamins, and nucleotides [31], allow for a faster growth rate, and are often used when changes to the standard conditions might restrict growth due to a requirement for one or more additional nutrients. For example, when *E. coli* is grown at or above 42 °C, it requires methionine [32]. Using this medium, *E. coli* can be maintained at a steady state growth (constant growth rate) for many generations (by dilution). Growth temperature, aeration state, pH, osmotic pressure, and growth phase are additional criteria for standardization [33].

3.1
Studying Microbial Organisms in Their Natural State is Usually not Amenable to 2D Gel Studies and These Conditions are Difficult to Mimic in a Laboratory Setting

Most pathogenic bacteria encounter two extremely different environments, life associated with its host organism and life in the often harsh and variable environment outside the animal host. Ideally, scientists would like to study the organism of interest in its natural state and environment. However, this is seldom possible. 2D gel studies have never been done on samples obtained from *E. coli* grown in an animal host due to difficulties in obtaining sufficient bacteria free from other contaminants. The proteins synthesized by *Salmonella* growing in macrophages revealed about 50 proteins whose expressed was up or down regulated during the course of the infection [34]. In this study, Abshire and Neidhardt compared the protein expression profiles obtained during growth of the bacteria in macrophages with protein expression profiles from cells stressed by treatment with polymyxcin B, heat shock, phosphate starvation, paraquat, acid shock, carbon starvation, sulfur starvation, peroxide, nitrogen starvation, or cold shock. The conclusion from this study was that the environment within the macrophage is not identical to any one of these single stress conditions.

Conditions that mimic life outside the animal host can to some degree be simulated in a laboratory. Although the shift from inside the host to outside the host often entails encountering many changes simultaneously, scientists have tended to simplify their studies by eliciting one or a small number of changes at a time. In addition, scientists have defaulted to developing defined media and determining growth conditions that allow the fastest growth rate of the organisms or conditions that allow a continuous growth rate for hundreds of generations. A good example of a study that attempted to mimic the transient from in the host (nutrient rich) to outside the host (nutrient starvation) was the work done in the 1980s on the bacteria *Vibrio* sp. S14 [35]. *Vibrio*, like many other marine bacteria, can survive prolonged starvation in the very dilute and nutrient-poor environment of the oceans. Many aspects of the physiology and biochemistry of the organism were monitored both during the shift-down, but also during recovery including morphological changes, cell wall, RNA and protein synthesis, ppGpp pool sizes, DNA synthesis, mean mRNA half-lives, stress resistance and changes in synthesis rates of individual proteins with 2D gels. Observations made in this study suggested that some aspects of the cell's programmed "shift-down" were similar to sporulation and germination observed in differentiating bacteria with two distinct difference. First, *Vibrio* did not become dormant. They maintained a low level of RNA and protein synthesis. Second, the recovery did not require a specific activation condition as does germination.

4 Proteome Analysis of Microbial Organisms Should Focus on Establishing the Goals of the Project

An important initial step is establishing the goals of the proteome project. Will the goal be to get biological evidence to support the interpretation of the genome analysis or to study the physiology of the organism? The frameworks for these two different types of studies, proteome mapping and global physiological analysis, are reviewed below.

4.1 Proteome Mapping

Several groups have focused on developing extensive proteome maps for different organisms [36–39]. In the 1980s Andrew Link and coworkers embarked on extensive proteome mapping project for *Escherichia coli*. They identified over 300 protein spots from 2D gels by -*N*-terminal sequencing [40]. Another proteome mapping effort for *E. coli* was published in 1996 [41] using a combination of analytical chemistry approaches. Proteome maps have also been published for many other organisms. Three of the more advanced proteome maps are for *Bacillus subtilis* [42], *Hemophilis influenza* [43], and the cyanobacterium *Synechocystis* sp. strain PCC6803 [44].

For the most part, proteome mapping is highly dependent on DNA sequence information to supply the data needed to identify proteins. However, Cordwell and Humphery-Smith [45] described and tested methods to identify proteins from organisms whose genome sequence is not known [46]. In these cases the identity is based on the cross-species conservation of amino acid sequence and of the masses of peptides generated from proteolytic cleavage [47–51]. In a recent study 45 protein spots from *Ochrobactrum anthropi* were identified using this approach [52].

For proteome mapping one sample of the chosen organism is sufficient for many subsequent studies. A strategy currently used by most groups focused on proteome mapping is to fractionate the sample by several methods, and then use narrow range isoelectric focusing and SDS PAGE techniques to allow maximum loading of each polypeptide species. In 1999 a proteome mapping project for *E. coli* was published that used hydroxyl appetite column fraction as an initial step in the analysis in order to identify less abundant proteins [53]. Many new proteins were identified in paper. Recent publications have suggested frameworks for cataloging data from proteome mapping studies [54, 55].

4.2
Global Physiological Studies

Projects focused on studying the physiology of a microbial organism through the expression of its proteins usually aims to provide physiological (reviewed in [2]), genetic, biochemical, regulatory, and architectural information for each cellular polypeptide. Physiological proteomics studies are more complex than for proteome mapping [56]. For these project hundreds and even thousands of different conditions must be analyzed:

- An example of how proteomic analysis has contributed to the genetics of an organism was the genetic mapping of the *rpoH* (*htpR*) gene using 2D gel analysis of expression from clones that complemented temperature-sensitive mutants in this gene [57, 58]. Using 2D gels Hirshfield et al. provided evidence that *E. coli* has two genes encoding lysyl-tRNA synthetase [59]. The genes were later found [60–62].
- There are several different ways that proteomic analysis has contributed to our understanding of the biochemistry and metabolism of an organism. In a couple of cases, 2D gel analysis has revealed that proteins being studied by different labs based on different properties or characteristics of the proteins were actually the same protein [63, 64]. Proteins needed for sulfur utilization in *Pseudomonas putida* -S-313 were identified using 2D gels. Matthews and Neidhardt identified that elevation of serine catabolism is one of the biochemical events resulting from heat shock in *E. coli* [65]. They correlated the effects of *metK* (and *lrp*) mutation (using strain RG62) on the heat shock response, the excretion of metabolites and incorporation of radioactive catabolic products of radiolabeled serine into other amino acids.
- Examples of how proteomics has contributed to the regulatory analysis of an organism are plentiful. The analysis of mutant in regulatory proteins has been a very effective method for identifying members of regulons. Most of the components of the heat shock regulon were identified by 2D gels [63, 66–68] in *E. coli*, but also in many organisms. The OxyR and many Lrp regulon members were identified by 2D gel analysis [69, 70].
- The very first 2D protein separation studies contributed to our understanding of the architectural organization of *E. coli*. Wittman's work on the 2D separa-

tion of ribosomes [71], and Link's work with membrane preparations [72] are excellent examples in this area. Another exciting application of 2D gels was done by Teixeira-Gomes et al. in a study designed to identify the presenting antigens (bacterial proteins) in an infection. They used serum from a ram naturally infected with *Brucella ovis* for immunoblotting studies to identify the immunogenic proteins recognized in the course of an infection [73].

4.3 There are Four Types of Information Collected in Proteomic Studies

Exploring the molecular physiology of an organism relies on the use of multiple scientific strategies and many methodologies. Proteomic studies provide information about the biochemical, genetic, regulatory, metabolic and architectural of the organism through diversely designed experiments, but so do other strategies and methods. How then can the data be brought together and organized so that the data can be probed [74]? This section describes an infrastructure upon which proteomic data can be organized into different "databases". The experiments done to construct each database yield different types of data on the cellular proteins. The databases are the reference 2D gel images that link data of a particular type together and can be linked together but can also be linked to data in other database (e.g., SWISS-PROT, GenBank).

1. *Theoretical data.* Many properties of proteins can be predicted from the DNA sequence of the genes that encode the proteins. From the deduced amino acid sequence, the pI and MW of a protein can be calculated. Cavalcoli published a study summarizing the number of proteins in different pI and MW ranges for *E. coli* [75]. This publication presented the limits of speculating the identity of a protein based on its migration on 2D gels. Post-translational modification sites can also be predicted. Within the annotation found in Swiss-Prot is the predicted and known post-translation modification sites. Cavalcoli has also created a database of all *E. coli* proteins that contains all the potential post-translational forms (J. Cavalcoli, unpublished data). In addition, Karlin has recently published a series of articles on a method to predict the abundance of proteins [76]. Collado-Vides and coworkers has developed a method for predicting the promoter structures of genes and thereby provides potential information about when the gene might be expressed [77].
2. *Genome expression data.* This work validates that an open reading frame can be transcribed and translated into a product. All of the efforts of proteome mapping contribute genome expression data. Another approach to obtaining genome expression data has been underway for 20 years. The approach is to express specifically the proteins cloned into a plasmid or phage. In 1976 Clarke and Carbon developed a colony bank containing segments of the *E. coli* chromosome [78]. In 1979 the first method for specific expression of plasmid-encoded genes was published, the maxi cell method [79]. In this method the *E. coli* chromosome is damaged by ultraviolet light and the cells are treated overnight with cycloserine. The chromosome-less cells can still transcript and translate proteins encoded by the high copy plasmid which survived the treat-

ment. A minicell method was also described [80]. This method is based on an *E. coli* mutant in which cell division occurs aberrantly resulting in the generation of minicells that have no chromosome, but does contain small plasmids and the cellular molecules needed for transcription and translation of the plasmid-encoded proteins. A year later another method was introduced which showed that plasmid-encoded genes were specifically transcribed and translation when cells were recovering from growth inhibition caused by chloramphenicol treatment [81]. Although all of these methods worked for some proteins, none allowed the expression of all the proteins on the clones. The first gene-protein index for *E. coli* was published in 1983 and included identifications based on these methods [82].

In 1987 a phage library containing almost the entire *E. coli* chromosome in approximately 20 kilobase segments was published [83]. The methods described above were tried with these phage clones [84], but also the T7 transcription system was developed [85]. Neidhardt's group has done extensive genome expression analysis with these Kohara clones by moving the inserts for the Kohara phages into a low copy plasmid containing T7 and SP6 promoters to drive transcription of the cloned segments ([86] and Clarke et. al. manuscripts in preparation). These results are being added to the protein expression information in Eco2Dbase.

RNA expression profiling (DNA chips and microarrays) is a way to extend the information in a genome expression map. The first global RNA analysis was published in 1993 [87]. The studies tested in this report were experiments that had previously been done with 2D gels. The correlation was good. For example, transcripts for all of the heat shock proteins were detected and several additional heat shock genes were revealed [87]. The DNA used in this first study were large segments of the genome. Membranes containing all the *E. coli* open reading frames are commercially available (Sigma Genosys, Inc.) and an *E. coli* genome array (Affymetrix, Inc.) is also commercially available. Such transcriptome studies will likely contribute to bacterial response-regulation data as well (see section below).

3. *Cellular abundance and architecture data.* When cells are grown in steady state conditions, proteomic analysis can determine the abundance or relative abundance of individual proteins. Subcellular fractionation can also be done to gain information about the cellular location and macromolecular arrangement of proteins.
   - *Abundance measurements.* Two of the large scale protein surveys done in the 1970s cataloged the levels of individual proteins during steady state conditions using different carbon sources and different growth temperatures [13, 88]. These studies allowed the grouping of proteins based on similarity of changes in their level in samples of cells grown in different conditions. These groupings in some cases allowed predictions of the potential function of the proteins. One set included the identified ribosomal proteins and other proteins that fit so tightly into the group that they were speculated to be ribosomal proteins (also based on their pI and MW). Another set, called the Ia3 proteins, contained no identified proteins at the time of the publication. Since then several of these proteins have been identified as enzymes in the

TCA cycle or enzymes in a precursory step to the TCA cycle. Information from these types of studies could be of enormous benefit for understanding the coordinated activities of cells.

As stated earlier in the introduction, one technical problem with such 2D gel studies is that many more proteins can be monitored in abundance studies than can be identified. A recent article reported the identification of many new proteins made possible by a fractionation step of the whole cell extract prior to 2D gel analysis [53]. The enrichment of the low abundance proteins made it possible to use traditional analytical methods to identify many proteins that had not been identified using whole cell extracts. Currently, the tracking of the proteins in the subfractions complicated the determination of the abundance of the proteins. Technically, this issue should be solvable with good computer-aided image analysis systems.

- *Architectural information.* As described above, there are technical problems with the study of macromolecular structures such as the ribosome, membranes, etc. Two difficulties with this approach are the lack of reproducibility of many fractionation methods (obtaining consistent quantitative amounts of each polypeptide) and the concern that some proteins may copurify with a macromolecule based on the purification method rather than on the association of the protein with that macromolecule. A classical example of the latter case is with the GroEL protein. Subramanian first found this protein in a preparation of ribosomes. He called the protein the A-protein, and proposed that it was a new factor involved in translation. In1980, Subramanian and others showed that the A-protein and the GroEL protein were in fact the same polypeptide [63]. Because the MW of the native GroEL is similar to that of the ribosome, the co-purification was thought to be an artifact. However, its role as a major cellular chaperone has sparked ideas that it may in fact be associated with the ribosome. The GroEL protein is a 14-mer so in a sucrose gradient, its sedimentation coefficient was similar to that of the ribosome. Whether GroEL really associates with ribosome in cells is still an open question.

The data in this abundance/architecture database is for defining steady state status of the cells and for discovering critical intracellular interactions. Such databases are also critical to the study of multi-cellular organisms. Determining what proteins are present in which cell types, tissues, and organs is important for understanding normal physiology as well as the pathology of diseases of these organisms. Norman and Leigh Anderson proposed a project to generate 2D gel maps of cells from nearly a hundred different tissues and organs of the human body, called the Human Protein Index [89]. Although this project was not funded at the level of the Human Genome Project, the information obtained from such an index will provide much insight and suggest many innovative chemotherapy and gene therapy strategies.

- *Response-regulation database.* Most of the 2D gel studies published on microbial organisms could be categorized as data for this type of map. In these projects proteins are grouped based on their response to a stimulus and suggested regulation pattern.

- *Response data.* Perhaps the most fascinating aspect of microorganisms is their ability to respond rapidly to changes in their environment. 2D gels have provided an effective method for monitoring the corresponding changes in proteins [57, 90]. In most cases pulse-chase labeling with radio-labeled amino acids are employed in these projects. This strategy permits one to view the synthesis rates of proteins during a particular period of time in the transition from one condition/state to another.

  Using pulse-chase experiments many more proteins can be detected than can be identified by analytical methods. However, these surveys should be thought of as the starting point of the analysis. Subsequent studies using a variety of genetic and biochemical approaches can yield the identity and the functions of the proteins. A perfect example of this is the discovery of the bacterial heat shock response using 2D gels. In the original publications, few of these proteins were identified and for those that did have names attached to them, they were only known because of the requirement for phage infection and reproduction [57, 91]. After the initial studies using 2D gels, literally hundreds of research labs began to focus on this small set of proteins found in bacterial but also found in all cells. Nearly 20 years later, the heat shock proteins are among the most well characterized proteins in biology.

  In addition to the response-regulation map being done for *E. coli*, an extensive response-regulation map has been developed for *B. subtilis* [42] and *V. cholera* [92]. This type of data has the most potential for antibacterial drug discovery and bacterial biotechnology projects.
- *Regulation data.* Proteomics has been useful in identifying members of regulatory networks as discussed in this section. Transcriptome studies should also be a very effective method for identifying members of regulons. Studies that combine transcriptome and proteomics analysis should reveal the predominate regulatory mode for individual genes, their products and activities of their products.

  For example, transcription factor sigma-32 has many different levels of control and therefore requires examination at multiple levels to provide an accurate picture of how it functions in the cell. The *rpoH* gene has at least four promoters that can trigger an increase in transcription of the gene under a variety of stress conditions [93]. The *rpoH* mRNA is also know to be transcribed at a higher rate than it is translated [38, 94] and thus there is much evidence that translational control plays a major role in determining when the product is synthesized. Third, the protein product, sigma-32 has a short half-life under some conditions [95] so, despite controls at the transcriptional and translational level, cells appear to be capable of controlling the level of this protein in the cell post-translationally. A fourth level of control of this protein exists. This protein acts as part of a macromolecule, RNA polymerase, to direct the initiation of transcription of about 20 genes [96, 97]. It is one of six sigma factors that have been found in *E. coli* and some evidence exists that the affinity of sigma-32 for core RNA polymerase may be higher than that of the "normal" sigma factor, sigma-70 [97]. This evidence predicts that the transcriptional, translational, and post-translational regulation are key to ensuring that this system is appropriately regulated and yields the proper level of the cells' chaperones and proteases most of which are transcriptionally controlled by this protein activity. Yet one study provides evidence that under chemostat conditions, sigma-32 levels are six to ten times higher than seen when cells are grown in standard laboratory conditions without concomitant high production of the products it transcriptionally regulates (J.N. D'Elia, A. Salyers, R.A. VanBogelen, unpublished results). Thus it appears that cells must also control the activity of sigma-32 presumably by controlling its interaction with core RNA polymerase [98]. Why has *E. coli* built in so much control of this one gene and its product? Is this unusual or are they many other genes/proteins whose regulation is this complex? Combine transcriptome and proteomic studies should reveal how genes/proteins are regulated.

4.4 The Ability to Track Proteins Through Many Experiments is the Key to Developing a Database of Proteomic Data

Because of the reproducibility of 2D gels, the idea of constructing a catalog of the data obtained from 2D gels emerged almost immediately after the technique was introduced. Spot names were assigned to the protein separated on 2D gels so that hundreds of unknown proteins could be tracked from gel to gel and from experiment to experiment [99]. Each new publication was tied to earlier publications through the spot names. For some catalogs, a master image was marked with all the spot names [100–102]. Others placed a grid over the master image so that X and Y coordinates could be assigned to each spot [82]. These spot names became accepted protein names, even appearing in the titles of publications [63, 68].

In addition to the experiments these groups were publishing, most had research efforts dedicated to identifying the protein spots to known proteins. For example, Neidhardt's group had a series of publication that correlated the spot names with known proteins [102, 103]. The early methods for identification of proteins relied heavily on the use of purified proteins (obtained from other investigators) that were comigrated on gels with whole cell extracts and also on antibodies to specific proteins that were used on immunoblots of the gels.

In the 1980s Celis organized the publication of a book that focused on the 2D gel methods and the catalogs being developed [104]. In 1990, 1991, and 1992 the journal Electrophoresis dedicated one issue a year to the publication of manuscripts that tied together the protein identifications and experimental data on the proteins. Also during this period many of these groups posted electronic versions of these databases at public sites, so that others investigators could download the entire database. Later web sites were developed. Models for 2D gel databases have been published for the gene to protein spot linkage (federated 2D gel database) but not for the tracking of proteins through experiments. Interestingly, most of the current web sites report only the protein identifications, but not the experimental data. The commercial value of these experimental data has been realized and many are available by subscription

(Proteome Sciences plc; Proteome, Inc.) or through contractile collaborations (Large Scale Proteomics). Robust databases have been developed by the proteomics companies mentioned above. These databases often have added direct linkage to the images, statistical analysis of the data, routine query and data visualization tools.

5 One of the Best Opportunities for Viewing what Happens to the Expression of Cellular Proteins when a Stimulus (Environmental Change, Chemical Insult, Mutant Strain, or the Under or Over Expression of a Gene) is Invoked on the Cells is with 2D Gels

There are hundreds of publications each year that include 2D gels to show how the synthesis rates or levels of proteins change in response to some stimulus. In the early 1980s Neidhardt and Gottesman coined a new term, stimulon, to be used to describe the set of proteins whose expression changes in a stimulus-response study [105]. This word has been used extensively by those doing proteomics because it is non-categorical with respect to the known genetics and biochemistry of the responding proteins.

Whether the data is qualitative (by gel gazing) or quantitative (by image analysis) most of the published 2D gel studies of a single condition compared to a control can be categorized into one of the two following specific aims. First aim, use the global approach to find one protein or a small number of proteins to be further explored by more reductionist approaches with a goal of making a connection between the "condition" and the function of the protein(s). Second aim, determine the breadth of the cell's response to the condition by analyzing both the *number* of proteins whose level or synthesis rate changes and the *amount* of cell's protein synthetic capacity is diverted toward a response to the condition.

5.1 Comprehensive Stimulus-Response Studies May Reveal Proteins that Provide a Function Needed by Cells in the "Stimulus" Condition

5.1.1 *Induced Proteins*

When the aim is to find an interesting protein whose function might relate to the condition of interest, most investigators pick the protein(s) whose level or synthesis increases most dramatically. The expectation is that the dramatic change in the protein is indicative of the importance of that protein in the cell's response. There are numerous examples of the success of this approach. The heat shock proteins are perhaps the best example. The synthesis rate (and in some cases the level) of these proteins is dramatically increased when the temperature at which the organism is grown approaches the fatal temperature for that organism. When these findings were first revealed [106] the functions of these proteins was not known. In fact, chaperones were a proposed but unidentified protein type.

5.1.2 *Repressed Proteins*

By focusing on proteins that are induced, have we missed important details about how cells cope with new conditions? Clearly the induced proteins correlate with new functions the cell needs. What about proteins whose synthesis rate or level decreases with a particular condition? Here are two reasons why decreases in the synthesis rates or levels of proteins may reveal interesting information about the physiology of the cells. First, a decrease may reflect that the cell no longer needs the protein's function, and thus the protein's synthesis rate or level is actively (by transcriptional repression, proteolysis, or post-translational modification) or inactively (by dilution with each generation) decreased. Second, a decrease could also indicate that the cells are no longer able to maintain the protein (by proteolysis or leakage of the protein from the cell) and this loss contributes to a secondary stress on the cells. A nice example of this scenario is the report from Steiner's group [107]. Proteomics was used to investigate the nature of the renal failure that is often an undesirable side in transplant patients receiving cyclosporine A. It was discovered that calcium binding protein, calbindin-D 28 KDa decreased in abundance in the kidneys of rats treated with cyclosporine A. The decrease of this protein was associated with the urinary calcium wasting and intratubular corticomedullary calcification in the kidney. Further studies showed that the level of this protein did not decrease in the kidneys of dogs and monkey that do not have nephrotoxic side affects of cyclosporine, but that the protein did decrease in levels in humans.

5.1.3 *Unresponsive Proteins*

What about the third category of response – the "no change" category – which is the proteins whose expression is not altered by a condition? No report has focused on this category although cell behavior could be learned by monitoring what proteins do not change. What can be learned by knowing the expression of protein is not affected by a stimulus. If a metabolic pathway or macromolecule is inhibited by an agent, a first hypothesis for the cell's response to the inhibitor would be an up-regulation of the protein(s) that is inhibited. In fact, in some cases the cells trigger others systems (e.g., stress systems).

5.2 Comprehensive Stimulus-Response Studies Can Be Focused More Intently on the Dynamics of the Cells' Response Rather than Specific Functions of the Proteins Induced or Repressed

In some instances the goal is not to gain insight into new or lost functions in the new condition. Instead, the goal is to gain an appreciation for the commitment cells make in order to adapt or survive in a new condition or a different growth or development phase of the organism. To date, very few studies with this focus have been published. One example is the study done on the developmental phases of *Streptomyces coelicolor* [25]. Most studies of *Streptomyces* focus on the production of antibiotics (with potential commercial value) which typically occurs in the transition phase of development. In this study 2D gels were used to monitor the levels of proteins as the organism moved through the different developmental phases. No protein identifications are reported in this report. A variety of multivariant statistical analysis methods were applied to the data (numerical data obtained from a computer-aided image analysis system) in order to gain new insight into *Streptomyces* physiology. Sixteen time points were examined (triplicate gels). Correlation analysis, cluster analysis, and principle component analysis resulted in the grouping of the timepoints into four groups (one timepoint was an outlier to all others). These groups corresponded with the four developmental phases previously characterized by morphological, physiological or biochemical characteristics.

6
Collation of Multiple Stimulus-Response Studies Often Reveals the Cause of the Change in Expression of a Protein

In a single stimulus-response study, is there really just a single change in the biology of the organism? Although it is true that the scientist can control the parameters of the experiment to ensure that a single stimulus is invoked, cells are robust and adaptive. In most cases, many changes in the cell's biology occurs with a single change in the condition the cells are grown [33]. For example, consider a simple case of steady state growth of bacteria in two cultures each containing different carbon sources, glucose and glycerol. Metabolism would be the major area of change. Some proteins will be present at higher or lower levels based on their roles in uptake, catabolism, and metabolism required for the particular carbon source. However, these are not the only changes that should be anticipated. As a consequence of differences in metabolism and energy output there would be differences in the rate of growth of the two cultures. The protein synthetic needs and the number of chromosome replication origins are also adjusted to the growth rate. Thus, the levels of the proteins involved in these two macromolecular synthesis pathways would also be expected to be different in these two conditions. For well-characterized proteins, like ribosome proteins, one can speculate that expression changes are due to a particular physiological parameter (growth rate), but for many proteins an obvious correlation may not be as easy.

One approach to identify correlations between protein expression and biological parameters is to integrate carefully information about the biology of the organism with the experimental data [29]. In the design of the experiment one should consider which parameters could be monitored while the experiment is underway? A second approach is to survey many conditions. For example, in one study *E. coli* was grown in steady state conditions with five different media compositions but all with the same growth temperature, same state of aeration, etc. [99]. Another study used the same growth media, but seven different growth temperatures [15]. With these combined studies, Neidhardt's group was able to identify proteins that responded to specific nutrient or environment conditions. For example, although 22 proteins were present at their highest level in acetate minimal media, only 5 were scored as uniquely elevated in acetate. The others were shown to become present at progressively higher levels as the generation time increased (acetate yielded the longest generation time). Another example, two proteins, called thermometer proteins, were found that appear to respond specifically to changes in growth temperature but do not exhibit a change in growth rate due to changes in nutrients supplied to the culture. Other proteins were found to be present at a specific level depending on the growth rate, regardless of media or temperature.

When comparing protein expression patterns obtained during a flux between steady state conditions, the question of which proteins are responding specifically and uniquely to the condition of interest becomes even more of a challenge. Comparisons of the proteins induced by different stress conditions have been made for a number of organisms.

Escherichia coli can grow on a variety of phosphorus sources. When phosphate is present in sufficient quantities, the growth rate of the organism is limited by factors other than the phosphorus source. The cellular response of *E. coli* to phosphate starvation is one of the best studied starvations. Nearly 30 years of genetic and biochemical studies have revealed many mechanistic details of the regulation that controls the metabolism of phosphate and other phosphorus sources. A comprehensive proteomics study of how cells respond to a depletion of phosphate as the sole phosphorus source (study of a transition) has been published [108] and compares this response to the adaptation of cells to growth on a limiting phosphorus source, phosphonate (a steady-state condition). Nearly 900 proteins were monitored in this study. The study offered some statistical information about the variation seen between identical experiments and gels from identical samples. A 13-page table gives the numerical data from the experiment. The analysis starts simply by presenting a histogram of the numbers of proteins induced and repressed in the study. A more complex distribution diagram allowed the overall responses to be compared by tracking the proteins. A Venn diagram was used to compare the proteins that respond to phosphate limitation to the proteins present at higher or lower levels in the phosphorus restrictive condition (phosphonate). The gene members of the PHO regulon are known to be induced by phosphate limitation and their protein products present at higher levels in phosphonate cultures. Thus, it was hypothesized that the overlap in the Venn diagram should also include members of the PHO regulon. Thus, this proteomics study suggested that the regulon may include 137 proteins. This number included 19 proteins that are repressed by both conditions, the first evidence PhoB may act as both transcriptional activators and repressors. A consensus promoter for members of the PHO regulons has been identified based on 10 promoters (over 30 protein products). The final figure in this paper presents important information that can only be gleaned from proteomics studies. These pie graphs illustrated the protein synthetic commitment *E. coli* allocated to phosphorus starvation.

Phosphate, nitrogen, and carbon starvations were done independently and as a triple starvation in *Vibrio vulnificus* [93]. The synthesis rate of proteins was measured (by 2D gel analysis) after 1 h of starvation. The results were presented as a Venn diagram showing the overlap in the protein responses. For example, of the 34 proteins induced by carbon starvation all were also induced by the triple starvation, whereas only 9 and 3 were also induced by nitrogen and phosphate starvation as a single starvation. A similar study in *B. subtilis* included carbon and nitrogen starvation in addition to heat, salt, ethanol, and oxidative stress [24]. In this study proteins were categorized as specific stress proteins (10 proteins) and general stress proteins (31 proteins). *Listeria monocytogenes* was monitored during acid, alkaline, SDS, deoxycholate, ethanol, heat and cold stress [109]. The levels of 68 proteins expressed in the control but elevated by one of more of these conditions was presented in a table. Another 57 proteins that had not been detected in the control were observed in one or more of these conditions. In *Listeria* no single protein was induced by all of these conditions, and unlike *B. subtilis*, most proteins responded uniquely to one condition tested in this study [110].

7 Relating Protein Responses to Physiology Allows for the Diagnosis of Cellular States of the Organism

Changes in protein expression (observed by the use of 2D gel electrophoresis) can be used to validate and extend information learned from other strategies and methods. This is an important application of proteomics. However, while exploring how *E. coli* alters its protein expression by numerous conditions, it was noted that predictions about the physiological state of the cells could be made [4]. At first, these observations seemed too obvious to declare as a significant scientific finding. However, while using proteomics to assist with drug discovery, this predictive capability proved to be more and more intriguing. The word signature is a common term use in science when interesting correlations are observed. Several examples of proteomics signatures have been presented [4].

Finding proteomics signatures requires that a large number of conditions have been studied, the proteins that respond uniquely to one condition have a high probability of being involved in adapting to or surviving the condition. This key question requires a transition in the analysis of the data. By surveying conditions and cataloging the resulting changes in protein expression, a comprehensive database can be built. To begin using this data to address questions in biology, a knowledge base of the organism and of the conditions used must be infiltrated with the protein data. With these combined data sets, correlations between the conditions tested and the protein responses (for a particular organism) can be identified. This step serves two purposes. First, it validates the protein expression data in the context of previous knowledge (and vice versa). Second, it also allows us to begin to learn about the cell behavior. The correlations become the framework for making predictions about the physiology of the organism using protein expression data.

The cellular sensor for stimulus-response networks is often the most difficult entity to identify. For the heat shock response network the sensor was especially difficult because of the large variety of agents that had been reported to induce the synthesis of this set of proteins. Some investigators have renamed the heat shock proteins as universal stress proteins. Conditions that truly mimicked a heat shock (defined by the induction and repression of the large sets of proteins that can be monitored by 2D gels) are treatment with antibiotics that target the bacterial ribosome [110]. It was found that all antibiotics that target the ribosome elicit either a heat shock response or a cold shock response. The level of induction of the heat or cold shock proteins are altered with increasing or decreasing temperature and the same was found to be true with antibiotics. As the concentration of the antibiotic was increased, the synthesis rate of the heat or cold shock proteins increased. The hypothesis drawn from this data was that the ribosome was the sensor for the heat and cold shock responses. It was proposed that the confirmation of the ribosome could be in the heat shock or cold-shock state and that these two states were exclusive. One experiment showed that the ribosome in the cold shock state (caused by treatment with tetracycline) blocked the normal heat shock response after a temperature shift. Although this article focused on the evidence that the ribosome was the sensor for the bacterial heat shock response, the finding that an environmental stress condition and a drug treatment could cause the same cell behavior (changes in the protein expression profile) was also a key finding for the authors.

Caution must be used in making these correlations because proteins are very dynamic cellular molecules and changes in their synthesis rates, levels and activity can be quickly altered by the cells' regulatory mechanism. In the mid-1980s a series of experiments were done and the resulting correlations were used to formulate an incorrect hypothesis. Initially the following set of observations were made:

1. Oxidative stress was found to induce a set of proteins that were all regulated by the OxyR protein [64]
2. Oxidative stress was found to increase the level of adenylated nucleotides [111]
3. Heat shock was found to increase the level of the same adenylated nucleotides [112] and the heat shock proteins
4. One agent, CdCl2, causing oxidative stress induces oxidative stress proteins and also the heat shock proteins [113]
5. Oxidative stress renders cells thermotolerant (increase survival to a shift to a lethal temperature) suggesting that the physiological role of the induction of the heat shock proteins was to protect the cells death at lethal temperatures [112]

These observations lead to a hypothesis that the two regulons, heat shock and oxidative stress, were activated by the same signal (the adenylated nucleotides) and were interrelated in their physiological role for cells. A more comprehensive study was done to compare directly the kinetics of induction of different stress responses and accumulation of nucleotides [114] in side-by-side studies of seven agents or conditions that elicited the induction of the stress responses. This study presented evidence that the hypothesis was incorrect. Specifically, it was shown that the heat shock regulon and OxyR regulon could be independently induced. H_2O_2 and ACDQ induced the OxyR regulon but not the heat shock regulon and shifts to 42 °C and treatment with ethanol cause an induction of the heat shock regulon but not the OxyR regulon. It was also demonstrated that the adenylated nucleotides did not accumulate in all conditions that induced the heat shock proteins suggesting that these nucleotides were not the inducing signal for this regulon. Finally, it was shown the induction of the heat shock proteins was not sufficient to render cells thermotolerant [115].

Another example of using correlations between protein expression and cell physiology had a more positive outcome – although even in this study the correlation could not be proven experimentally. In 1993 Abshire and Neidhardt [116] used correlations between the degree of post-translation modification of a proteins and growth rate in vivo to estimate the growth rate of cells growing intracellularly. The primary aim of the study was to try to identify the stresses the bacteria were encountering in the macrophage [34], but they also used the protein expression data to estimate the generation time of the cells while in the macrophages. They used a protein post-translational modification as one indicator of growth rate. The ribosomal protein, L12, can be modified (acetylation) in both *E. coli* and *Salmonella*. Previously, a correlation between growth rate and the ratio of modified protein to total L12 protein had been found in *E. coli*. The same correlation was found to exist in *Salmonella*. The degree of modification of this protein in the bacterial cells growing intracellularly estimated a growth rate that was much faster than the rate proposed by plating the bacteria after disruption of the macrophage cells. Results from other experiments suggested that the degree of modification of L12 was the more accurate estimator of generation time.

A recent report in Science demonstrated that mRNA expression studies could also be used as a molecular diagnostic tool [117]. In this study 38 acute leukemia samples were analyzed using DNA microarrays. The expression of nearly 7000 genes was monitored. They developed a method called neighborhood analysis to identify a set of genes whose expression highly correlated with the classification of the cancer (acute myeloid leukemia or acute lymphoblastic leukemia). A set of 50 genes was chosen as the distinguishing predictor set to diagnosis the type of cancer. The expression of these genes was then monitored in an additional 34 leukemia samples. The predictor set was effective on 29 of the 34 samples with 100% accuracy according to other clinical indicators.

These examples demonstrate the potential for using protein (and mRNA) expression data as biotechnology tools for the molecular diagnosis [118]. For basic research studies these tools will focus on providing more details of the molecular physiology of biological system. For clinical research, these tools will help in the development of more reliable indicators of disease so more successful treatments can be used [119].

8
The Wealth of Biological Data Being Generated is Providing the Impetus for a Wet Lab to Dry Lab Transition in Biological Studies

The next step in using protein (and RNA) expression data as diagnostic indicators and as monitors of cell behavior is to use them as test sets for formulating mathematical simulations of how cells work. The value of cell models and simulations is to allow scientists to do a "dry" experiment (using a computer) before the real, "wet" experiment is done. By simulating a cellular response, the scientist can look at many parameters in the cell and can predict which parameters could simultaneously be monitored in the real experiment. Better experimental designs mean better experimental outcomes. A variety of models and simulations are being constructed to describe different aspects of "biological life"; however, to date no one has integrated these functional genomics studies into any models.

Cellular metabolism is one area where modeling and simulations are being made and used [120–128]. Two reasons for the intense interest in modeling metabolism are: first, many details of enzymes are known from intensive biochemical studies; and second, these models have immediate application in industry where the re-engineering of biological systems has the potential to yield commercially valuable material. There are numerous success stories where genetic engineering of an organism has yielded the desired outcome (reviewed in [129]). However, other attempts have failed. For example, yeast strains were genetically engineered to over express the enzymes in the glycolysis pathway with the expected outcome of increased ethanol production [130]. However, no significant increase was obtained. Another example of frequent genetic engineering failures is with the overproduction of foreign proteins in *E. coli*. Some proteins are produced in high quantity and have activity, whereas other proteins are inactive, insoluble, and/or just poorly expressed. What cell behaviors jeopardize these efforts? Can these behaviors be controlled? James E. Bailey wrote a commentary for Nature Biotechnology suggesting that protein and RNA expression data might be very beneficial for overcoming some of the current difficulties with metabolic engineering [131].

Regulatory circuitry is an area where simulations and modeling has been done [132–137]. Interesting developments in this area include the application of the electrical engineering simulation program, SPICE to modeling biological regulatory circuits (http://www.swiss.ai.mit.edu/~rweiss/bio-programming/). In November of 1999 it was announced that a group of labs in the United States would work together to build a map of cell signaling that would be the impetus for building a "virtual cell" [138]. This effort called the Alliance for Cellular Signaling (AFCS), headed by Alfred Gilman, brings together systems engineers, biologists, and information scientists and will cost about 10 million dollars a year to run [138]. The goal is that the products of this effort be publicly available to the scientific community.

There are other efforts required for modeling whole systems [139]. For example, the E-cell project is structured to include eventually interrelated models of all aspects of *E. coli* biology [140]. A commercial effort to model the human disorders obesity and asthma is also underway (Entelos, Inc.; http://www. Entelos, com). This recent burst of effort toward building simulations and modeling is likely only the beginning. All of the data being generated by genome and functional genomics has clearly spurred this effort toward the ultimate goal in biology that is to be able to "solve the cell with mathematics".

9 Concluding Remarks

This chapter has reviewed work contributing to the discovery phase of applying proteomics to elucidate the molecular physiology of the microbial organism,

Escherichia coli. These concluding remarks are a vision for proteomics in the twenty-first century. Suppose the year is 2015 – could this be an accurate description of the current state of proteomics technologies and how proteomics has been integrated with other approaches to create the current state of biological research on microorganisms?

9.1 Technical Improvements for Proteomics

9.1.1 *Detection and Identification of Proteins*

Of the predicted open reading frames in a genome, 80% can be detected (quantitatively) and linked to the encoding gene. The four critical criteria for detection of proteins are isoelectric point, mass, abundance, and solubility of the proteins. The most difficult single criterion is the detection and identification of low abundance proteins. The distribution of the proteins in a complex cellular sample is a one-tailed extreme distribution that extends over eight orders of magnitude.

9.1.2 *Reproducibility of the Separation System*

A method of protein separation that yields a 99% success rate is standard in proteomics labs.

9.1.3 *Image Analysis*

Image analysis methods process and search images at a rate equivalent to the throughput of the separation system. The typical lab generates and processes 100 image per day per person.

9.1.4 *Data Analysis and Data Mining*

Standard routines for data analysis that include quality control checks and determination of reliability are implemented. A series of data mining routines that relate the data to previous knowledge and explores the data for new information are developed, standardized, and implemented.

9.2 Dynamic Cellular Parameters Measured in Routine Analysis

9.2.1 *Measurement of Cellular Proteins*

Proteomics has proven to be a very powerful approach to reveal novel information about biology and to diagnose the states of cells, but it became even more powerful when done in concert with other approaches. Labs routinely monitor the level, the synthesis rate, and the degradation rate of each protein. In addition all post-translation species of the protein are identified and monitored. In addition, the mechanism of regulation of activity at the transcriptional, translational, and post-translational level is being revealed for each and every gene.

9.2.2
Measurements of mRNA Molecules

Determination of the level and ideally the synthesis rate of each mRNA molecule are done using samples from the same culture used for proteomic studies. The wealth of protein and RNA expression profiles has revealed interesting rules for relationships between the messenger and the product.

9.2.3
Macromolecule and Cell Structure Determinations

Reproducible and biologically accurate methods for isolation of many macromolecules have been established. This information has had an impact on reconstructing the architecture of cells. A diverse set of approaches was established to detect and follow structure in cells. Action movies of the structures are available (using methods like reconstructive electron microscopy) so that processes such as cell division, chromosome segregation, protein translation, protein translocation, and chemotaxis can be viewed "in action".

9.3
Establishing the Molecular Physiology Base

9.3.1
Development of an Intensive Bioinformatics Databases of the DNA Sequence

This includes the prediction of what the organism is capable of doing, what conditions it can grow in, and which conditions it can survive. Metabolic potential, growth conditions, and stress responses can be predicted from genome information. The database focuses on collecting experimental data that is used to determine if all of the predictions from bioinformatics are correct. A critical step is a clear determination and definition for steady state growth.

9.3.2
Completion of the Genome Expression Database

The experimental determination that predicted open reading frames encoding a protein is important. This is determined by a high throughput expression system that allows the actual ORFs to be produced and analyzed. This system overrides the gene's own transcriptional and translational regulation.

9.4 Multivariant and High-Density Genomic Exploration

9.4.1 *Libraries of Conditional Mutants were Constructed*

Mutants in each gene on the chromosome are available and controllable so that a determination of how cells adapt (or do not adapt) to the absence of the gene's product can be determined by the dynamic cellular parameters listed above. Because bioinformatics has predicted when the gene is necessary, the number of experiments can be constrained.

9.4.2 *Exploration of the Transition Phase Between Two Steady-State Conditions*

Experiments continue to be done to analyze the dynamic cellular parameters during the transition from one steady-state condition to another where the condition is defined as a change in the cells' environment or genetic state (see above). The time requirements for the transition is experimentally determined and defined.

9.4.3 *Exploration of the Sequence of Events Elicited by Terminal Conditions*

Many conditions (both environmental and genetic) are bacteriostatic or bactericidal. The exploration and analysis of the dynamic cellular parameters in these conditions is revealing how the cells attempt to counteract, evade and survive these conditions. The terminal state of cells is being defined.

9.5 Interconnected Mathematically Descriptive Models and Simulation

9.5.1 *Models of the Structural Features of Cells*

The structural features of cells have been monitored dynamically in vivo for many conditions. This monitoring being done in concert with extensive monitoring means the physicochemical parameters of intact cells have led to many new findings in the area of physical chemistry. A new discipline, molecular thermodynamics, has emerged and is one of the "hottest" areas of science. This discipline defines the chemistry outside dilute solutions and has transformed the framework upon which mathematical models of the infrastructure of the cells are built.

9.5.2 *Models and Simulations of the Processes the Cells Can Perform*

All the major processes of the cell have been modeled as independent systems. Metabolic models are now open system models that integrate active and passive movement of molecules in and out of cells and include data on the synthesis rate, degradation rate, level, and specific activity of enzymes. Two types of network connections have been modeled and made available as simulations. One type is the regulatory networks that use the database of gene-expression phenotypes of each individual gene (from transcriptome and proteome data) and the circuitry systems (Bayesian networks) that interconnect the genes and predicts modes of regulation. The second type of network connection is the crosstalk systems that interrelate cellular process (metabolic, polymerization, and growth) with cellular structures (membranes, chromosomes, and ribosomes).

9.5.3 *Simulate the Performance of Cells in a Pseudo Ecological System*

Most simulations focus on applications in biotechnology and drug discovery. These simulations predict the best genetic and environment plan for using microorganisms for a desired fermentation outcome. Using these simulations, Streptomyces and yeast strains have become extremely important for human life. The DNA sequence of these organisms is determined "in the field", as is their potential for useful biotechnology applications. Thus patents are often filed before the organism arrives at the research or production laboratory. Both anti-bacterial and anti-viral antibiotic development has grown exponentially in the last decade. All antibiotics developed over the last decade are narrow spectrum not only with respect to causative agent, but also with respect to the tissue or cell type affected. Researchers have recognized that human cells create the environment for the infectious agent and thus influence which genes and gene products are required. Most diseases have been shown to stem from bacterial or viral infections either by triggering an autoimmune response or by horizontal gene transfer to human cells. In many cases, the infection occurred historically (some traced back more then ten generations) and has been latent for generations. Pharmacoecology studies provide much of the data for simulations used in drug discovery.

10 References

1. VanBogelen RA, Greis KD, Blumenthal RM, Tani TH, Matthews RG (2000) Trends Microbiol 7:320
2. VanBogelen RA, Neidhardt FC (1999) Preparation of *Escherchia coli* samples for 2-D gel analysis, 2-D proteome analysis protocols. In: Link A (ed) Methods in molecular biology, vol 112. Humana Press, chap 3
3. VanBogelen RA, Abshire KZ, Pertsemlidis A, Clark RL, Neidhardt FC (1996) Gene-protein database of *Escherichia coli*. In: *Escherichia coli* and *Salmonella typhimurium*: cellular and molecular biology, 6th edn. American Society for Microbiology, Washington, DC
4. VanBogelen RA, Schiller E, Thomas JD, Neidhardt FC (1999) Electrophoresis 20:2149
5. O'Farrell PH (1975) J Biol Chem 250:4007
6. Anderson NL, Anderson NG (1998) Electrophoresis 19:1853
7. Wasinger VC, Cordwell SJ, Cerpa-Poljak A, Yan JX, Gooley AA, Wilkins MR, Duncan MW, Harris R, Williams KL, Humphery-Smith I (1995) Electrophoresis 16:1090
8. Anderson L, Sellhamer J (1997) Electrophoresis 18:533
9. Gygi SP, Rochon PY, Franza BR, Aebersold R (1999) Mol Cell Biol 19:1720
10. VanBogelen RA, Olson ER (1995) Biotechnol Ann Rev 1:69

11. Hunkapiller MW, Hood LE (1983) Science 219:650
12. Shevchenko AJ, Podtelejnikov ON, Sagliocco AV, Eilm F, Vorm M (1996) Proc Natl Acad Sci USA 93:440
13. Haynes PA, Gygi SP, Figeys D, Aebersold R (1998) Electrophoresis 19:1862
14. Traini M, Gooley AA, Ou K, Wilkins MR, Tonella L, Canchez JC, Hochstrasser DF, Williams KL (1998) Electrophoresis 19:1941
15. Herendeen SL, VanBogelen RA, Neidhardt FC (1979) J Bacteriol 139:185
16. Walsh BJ, Molloy MP, Williams KL (1998) Electrophoresis 19:1883
17. Chevallet M, Santoni V, Poinas A, Rouquie D, Fuchs A, Kieffer S, Rossignol M, Lunardi J, Garin J, Rabilloud T (1998) Electrophoresis 19:1901
18. Gorg A, Obermaier C, Boguth G, Weiss W (1999) Electrophoresis 20:712
19. Gygi SP, Rist B, Gerber SA, Turecek F, Gelb MH, Aebersold R (1999) Nat. Biotechnol 17:994
20. Monardo PJ, Boutell T, Garrels JI, Latte GI (1994) Comput Appl Biosci 10:137
21. Smilansky Z (2110) Electrophoresis 22:1616
22. Steinberg TH, Lauber WM, Berggren K, Kemper C, Yue S, Patton WF (2000) Electrophoresis 21:497
23. Unlu M, Morgan ME, Minden JS (1997) Electrophoresis 18:2071
24. Antelmann H, Bernhardt J, Schmid R, Mach H, Volker U, Hecker M (1997) Electrophoresis 18:1451
25. Vohradsky J, Li XM, Thompson CJ (1997) Electrophoresis 18:1418
26. Puglia AM, Vohradsky MJ, Thompson CJ (1995) Mol Microbiol 17:737
27. Schmid H, Schmitter RD, Blum P, Miller M, Vonderschmitt D (1995) Electrophoresis 16:1961
28. Cooper S (2000) ASM News 66:71
29. Neidhardt FC (1999) J Bacteriol 181:7405
30. Neidhardt FC, Bloch PL, Smith DF (1974) J Bacteriol 119:736
31. Wanner BL, Kodaira R, Neidhardt FC (1977) J Bacteriol 130:212
32. Ron EZ, Davis BD (1971) J Bacteriol 107:391
33. Neidhardt FCJ, Ingraham L, Schaechter M (eds) (1990) Physiology of the bacterial cell: a molecular approach. Sinauer Associates, Sunderland MA
34. Abshire KZ, Neidhardt FC (1993) J Bacteriol 175:3734
35. Nystrom T, Albertson NH, Flardh K, Kjelleberg S (1990) FEMS Microbiol Ecol 74:120
36. Binz PA, Muller M, Walther M, Bienvenut WV, Gras R, Hoogland C, Bouchet G, Gasteiger E, Gabbretti R, Gay S, Palagi P, Wilkins MR, Rouge V, Tonella L, Paesano S, Rossellat G, Karmime A, Bairoch A, Sanchez JC, Appel RD, Hochstrasser DF (1999) Anal Chem 71:4981
37. Blackstock W, Weir MP (1999) Trends Biotechnol 17:121
38. Kamath-Loeb AS, Gross CA (1991) J Bacteriol 173:3904
39. Wilkins MR, Sanchez JC, Gooley AA, Humphrey-Smith I, Hochstrasser DF, Williams KL (1996) Genet Eng Rev 13:19
40. Link AJ, Robison K, Church GM (1991) Electrophoresis 18:1259
41. Pasquali C, Frutiger S, Wilkins MR, Hughes GJ, Appel RD, Bairoch A, Schaller D, Sanchez JC, Hochstrasser DF (1996) Electrophoresis 17:547
42. Bernhardt J, Buttner K, Scharf C, Hecker M (1999) Electrophoresis 20:2225
43. Langen H, Takacs B, Evers S, Berndt P, Lahm HW, Wipf B, Gray C, Fountoulakis M (2000) Electrophoresis 21:411
44. Sazuka T, Yamaguchi M, Ohara O (1999) Electrophoresis 20:2160
45. Cordwell SJ, Humphery-Smith I (1997) Electrophoresis 18:1410
46. Cordwell SJ, Basseal DJ, Humphery-Smith I (1997) Electrophoresis 18:1335
47. Henzel WJ, Billeci TM, Stults J, Wong SC, Grimley C, Watanabe C (1993) Proc Natl Acad Sci USA 90:5011
48. Mann M, Hojrup P, Roepstorff P (1993) Biol Mass Spectrom 22:338
49. Pellegrini M, Marcotte EM, Thompson MJ, Eisenberg D, Yeates TO (1999) Proc Natl Acad Sci USA 96:4285
50. Shaw G (1993) Proc Natl Acad Sci USA 90:5138

51. Yates JR III, Speicher S, Griffin PR, Hunkapiller T (1993) Anal Biochem 214:397
52. Wasinger VC, Urquhart BL, Humphery-Smith I (1999) Electrophoresis 20:2196
53. Fountoulakis M, Takacs MF, Berndt P, Langen H, Takacs B (1999) Electrophoresis 20:2181
54. Cordwell SJ, Nouwens AS, Verrills NM, McPherson JC, Hains PG, Van Dyk DD, Walsh BJ (2000) Electrophorsis (in press)
55. Mollenkopf HJ, Jungblut PR, Raupach B, Mattow J, Lamer S, Zimny-Arndt U, Schaible UE, Kaufmann SHE (1999) Electrophoresis 20(11):2172
56. Neidhardt FC, VanBogelen RA (2000) Proteomic analysis of bacterial stresss responses. In: Storz G, Hengge-Aronis R (eds) Bacterial stress response. American Society for Microbiology, Washington DC
57. Neidhardt FC, VanBogelen RA (1981) Biochem Biophy Res Comm 100:894
58. Neidhardt FC, VanBogelen RA, Lau ET (1983) J Bacteriol 153:597
59. Hirshfield IN, Bloch PL, VanBogelen RA, Neidhardt FC (1983) J Bacteriol 146:345
60. Clark RL, Neidhardt FC (1990) J Bacteriol 172:3237
61. Emmerich RV, Hirshfield IN (1987) J Bacteriol 169:5311
62. VanBogelen RA, Vaughn V, Neidhardt FC (1983) J Bacteriol 153:1066
63. Neidhardt FC, Phillips TA, VanBogelen RA, Smith MW, Georgalis Y, Subramanian AR (1981) J Bacteriol 145:513
64. VanBogelen RA, Hutton ME, Neidhardt FC (1990)Electophoresis 11:1131
65. Matthews RG, Neidhardt FC (1989) J Bacteriol 171:2619
66. Neidhardt FC, VanBogelen RA, Vaughn V (1984) Ann Rev Genet 18:295
67. Phillips TA, VanBogelen RA, Neidhardt FC (1984) J Bacteriol 159:283
68. Tilly K, VanBogelen RA, Georgeopoulos C, Neidhardt FC (1983) J Bacteriol 154:1505
69. Christman MF, Morgan RW, Jacobson FS, Ames BN (1985) Cell 41:753
70. Ernsting BR, Atkinson MR, Ninfa AJ, Matthews RG (1992) J Bacteriol 174:1109
71. Chersi A, Dzionara M, Donner D, Wittman HG (1968) Mol Gen Genet 101:82
72. Link AJ, Hayes LG, Carmack EB, Yates JR III (1997) Electrophoresis 18:1259
73. Teixeira-Gomes A, Cloeckaert PA, Bezard G, Bowden RA, Dubray G, Zygmunt MS (1997) Electrophoresis 18:1491
74. VanBogelen RA, Abshire KZ, Moldover B, Olson ER, Neidhardt FC (1997) Electrophoresis 18:1243
75. Cavalcoli JD, VanBogelen RA, Andrews PC, Moldover B (1997) Electrophoresis 18:2703
76. Karlin S, Mrazek J, Campbell AM (1998) Mol Microbiol 29:1341
77. Salgado H, Santos-Zavalenta A, Gama-Castro S, Millan-Zarate D, Blattner FR, Collado-Vides J (2000) Nucl Acid Res 28:65
78. Clarke L, Carbon J (1976) Cell 9:91
79. Sancar A, Hack AM, Rupp WD (1979) J Bacteriol 137:692
80. Reeve J (1979) Methods Enzymol 68:493
81. Neidhardt FC, Wirth R, Smith MW, VanBogelen RA (1980) J Bacteriol 143:535
82. Neidhardt FC, Vaughn V, Phillips TA, Bloch PL (1983) Microbiol Rev 47:231
83. Kohara I, Akiyama K, Isono K (1987) Cell 50:495
84. Neidhardt FC, Appleby DB, Sankar P, Hutton ME, Phillips TA (1989) Electrophoresis 10:116
85. Studier FW, Moffat BA (1986) J Mol Biol 189:113
86. Sankar P, Hutton ME, VanBogelen RA, Clark RL, Neidhardt FC (1993) J Bacteriol 175:5145
87. Chuang SE, Daniels DL, Blattner FR (1993) J Bacteriol 175:2026
88. Lemaux PG, Herendeen SL, Bloch PL, Neidhardt FC (1978) Cell 13:427
89. Anderson NG, Anderson NL (1981) JAMA 246:2620
90. VanBogelen RA, Neidhardt FC (1990) FEMS Microbiol Ecol 74:121
91. Yamamori T, Yura T (1982) Proc Natl Acad Sci USA 79:860
92. Ostling J, McDougald D, Marouga R, Kjelleberg S (1997) Electrophoresis 18:1441
93. Fujita N, Ishihama A (1987) Mol Gen Genet 210:10
94. Straus DB, Walter WA, Gross CA (1987) Nature 329:348
95. Nagai H, Yuzawa H, Yura T (1991) Proc Natl Acad Sci USA 88:10,515

96. Grossman AD, Erickson JW, Gross CA (1984) Cell 38:383
97. Landick R, Vaughn V, Lau ET, VanBogelen RA, Erickson JW, Neidhardt FC (1984) Cell 38:175
98. Liberek K, Galitski TP, Zylicz M, Georgopoulos C (1992) Proc Natl Acad Sci USA 89:3516
99. Pedersen S, Bloch PL, Reeh S, Neidhardt FC (1978) Cell 14:179
100. Bravo R, Celis JE (1980) J Cell Biol 84:795
101. Bravo R, Celis JE (1982) Clin Chem 28:766
102. Bloch PL, Phillips TA, Neidhardt FC (1980) J Bacteriol 141:1409
103. Phillips TA, Bloch PL, Neidhardt FC (1980) J Bacteriol 144:1024
104. Celis JE, Bravo R (eds) (1984) Two-dimensional gel electrophoresis of proteins: methods and applications. Academic Press, London New York
105. Gottesman S, Neidhardt FC (1983) Global control systems. In: Beckwith J, Davies J, Gallant JA (eds) Gene function in prokaryotes. Cold Spring Harbor Laboratory, Cold Spring Harbor, NY
106. Schlesinger MJ, Ashburner M, Tissieres A (eds) (1982) Heat shock from bacteria to man. Cold Spring Harbor Laboratory, Cold Spring Harbor, NY
107. Aicher LD, Wahl A, Arce GO, Steiner S (1998) Electrophoresis 19:1998
108. VanBogelen RA, Olson ER, Wanner BL, Neidhardt FC(1996) J Bacteriol 178:4344
109. Phan-Thanh L, Gormon T (1997) Electrophoresis 18:1464
110. VanBogelen RA, Neidhardt FC (1990) Proc Natl Acad Sci USA 87:5589
111. Bochner BR, Lee PC, Wilson SW, Cutter CW, Ames BN (1984) Cell 37:225
112. Lee PC, Bochner BR, Ames BN (1983) Proc Natl Acad Sci USA 80:7496
113. Courgeon AM, Maisonhaute C, Best-Belpomme M (1984) Cell Res 153:515
114. VanBogelen RA, Kelley PM, Neidhardt FC (1987) J Bacteriol 169:26
115. VanBogelen RA, Acton M, Neidhardt FC (1987) Genes Develop 1:535
116. Abshire KZ, Neidhardt FC (1993) J Bacteriol 175:3744
117. Golub TR, Slonim DK, Tamayo P, Huard C, Gassenbeek M, Mesirov JP, Coller H, Loh ML, Downing JR, Caligiuri MA, Bloomfield CD, Lander ES (1999) Science 286:531
118. Huang S (1999) J Mol Med 77:469
119. Gulino A (1999) Forum (Genova) 9:37
120. Edwards JS, Palsson BO (1998) Biotechnol Bioeng 58:162–169
121. Ehlde M, Zacchi G (1995) CABIOS 11:201
122. Hofmeyr JH (1986) Comput Appl Biosci 1:5
123. Liao JC (1993) Curr Opin Biotechnol 2:211
124. Mendes P (1997) Trends Biochem Sci 22:361
125. Palsson BO (1997) Nature Biotechnol 15:3
126. Sauer U, Lasko DR, Fiaux J, Hochuli M, Glaser R, Szyperski T, Wuthrich K, Bailey JE (1999) J Bacteriol 181:6679
127. Schilling CH, Edwards JS, Palsson BO (1999) Biotechnol Prog 3:288
128. Stephanopoulos G (1994) Curr Opin Biotechnol 2:196
129. El-Gewely MR (1995) Biotechnology domain. In: El-Gewely MR (ed) Biotechnology annual review, vol 1. Elsevier, Amsterdam Lausanne New York Oxford Shannon Tokoyo, pp 5–68
130. Schaaff I, Heinisch J, Zimmermann FK (1989) Yeast 5:285
131. Bailey JE (1999) Nat Biotechnol 17:616
132. Arkin A, Ross J, McAdams HH (1998) Genetics 149:1633
133. Bray D, Bourret RB, Simon MI (1993) Mol Biol Cell 4:469
134. Gillespie DT (1999) J Phy Chem 81:2340
135. McAdams HH, Arkin A (1997) Proc Natl Acad Sci USA 94:814
136. McAdams HH, Shapiro L (1995) Science 269:650
137. Meyers S, Friedland P (1984) Nucl Acid Res 12:1
138. Abbott A (1999) Nature 402:219
139. Reddy B, Yin J (1999) AIDS Res Human Retroviruses 15:273
140. Tomita MK, Hashimoto K, Takahashi TS, Shimizu Y, Matsuzaki F, Miyoshi K, Saito S, Tanida K, Yugi K, Venter JC, Hutchison CA III (1999) Bioinformatics 15:72

Received: April 2002

Adv Biochem Engin/Biotechnol (2003) 83: 57–92
DOI 10.1007/b11031

A Proteomic View of Cell Physiology of *Bacillus subtilis* – Bringing the Genome Sequence to Life

Michael Hecker

Ernst-Moritz-Arndt-Universität Greifswald, Institut für Mikrobiologie, F.-L.-Jahn-Strasse 15, 17487 Greifswald, Germany. *E-mail: hecker@biologie.uni-greifswald.de*

The genome sequence is the "blue-print of life", and the proteomic approach brings this genome sequence to life. Simple model systems are urgently required to "train" this transformation of the genome sequence into life: why not *Bacillus subtilis*, the model organism for Gram-positive bacteria and of functional genomics?

By combination of the highly sensitive 2D protein gel electrophoresis with the identification of the protein spots by microsequencing or mass spectrometry we established a 2D protein index of *Bacillus subtilis*. In order to depict the entire proteome of a *B. subtilis* cell, alkaline, cell-wall associated, or extracellular proteins were also included. The proteins of this database (see http://microbio2.biologie.uni-greifswald.de:8880/sub2d.htm) were allocated to proteins with house-keeping functions typical of growing cells and to proteins synthesized particularly in non-growing cells. A computer-aided evaluation of the 2D gels loaded with radioactively-labeled proteins from growing or stressed/starved cells proved to be a powerful tool for the analysis of global regulation of the expression of the entire genome. This is shown for the analysis of glycolysis/TCA cycle (house keeping proteins) and for the analysis of the heat stress stimulon.

For the heat stress stimulon it is demonstrated how the proteomic approach can be used: (i) to define the structure of a stimulon, (ii) to dissect stimulons into regulons, (iii) to analyze the regulation, structure, and function of unknown regulons, (iv) to define overlapping regulons or modulons, and finally (v) to explore complex adaptational networks.

Furthermore, it will be demonstrated how the "dual channel pattern comparison" [24] or "proteomics signature" (R. VanBogelen) can be used for a comprehensive understanding or prediction of the physiological state of growing or starving cell populations. This is shown for glucose-starved cells.

In order to describe the structure and function of gene regulation groups it is generally recommended to complement the proteomics approach with DNA array technologies. Further studies will focus on the analysis of the global regulation of gene expression by the proteomic approach that cannot be addressed by the application of DNA array techniques:

- The phosphoproteome and its implications in signal transduction
- The global control of protein stability
- Protein targeting and protein secretion

Keywords. Physiological proteomics, Transcriptomics, Reference map, Adaptational network, Protein targeting, Post-translational modification

1 Introduction – Bacteria as Model Systems for "Functional Genomics"

In 1995 a new era in biology was opened with the publication of the first genome sequence of a living organism, the bacterium *Haemophilus influenzae* [1]. For the first time the genome sequence of an organism became available, giving an opportunity for understanding a living cell and thus life in general. This new era has culminated for the moment in the publication of the human genome sequence in February 2001. The genome sequence, however, represents only the blue-print of a living cell, not "life itself". Now functional genomics is necessary to bring this genome sequence to life, in terms of genome-wide mRNA profiling, proteomics, or bioinformatics. The functional genomics of the human being will be of enormous importance, not only for analyzing processes of human development and differentiation, but also for a basic understanding of many diseases, leading to new developments in diagnostics and therapy. However, the "human being", with a huge number of different cell types, tissues, and organs, is much too complex to lead to a comprehensive understanding of life. A world-wide focus on a few model systems of human cell differentiation and disease is required for such a comprehensive study and for understanding developmental processes in terms of the global regulation of gene expression. Even if such model systems become available in the near future, the potential expression of 30,000 to 40,000 genes and of many more proteins as the main players of life, their structure and function, their interplay, and cellular destination would need to be taken into consideration. It follows that simple organisms are required as model systems to "ensure the transformation of the genome sequence into life".

Bacillus subtilis, a Gram-positive bacterium, has become a model organism for functional genomics through the joint efforts of many research groups in the

USA, Japan, and Europe [2]. There is extensive knowledge of the biochemistry and genetics of *B. subtilis* through the world-wide interest in *B. subtilis* as a simple model for understanding cell differentiation on a molecular scale. Furthermore, the long-standing use of members of the genus *Bacillus* in biotechnology has necessitated studies of the genetics and cell physiology of *Bacilli*. The *B. subtilis* genome sequence, published in 1997 [3] as the result of a joint research program in Europe and Japan, has revealed more than 4100 genes. Of these 4100 genes more than 1700 code for proteins whose functions are still unknown. The elucidation of the function of these proteins is a big challenge for future research. A start has been made in this direction, through the construction of a mutant library containing single mutations in each single gene of unknown function and a comprehensive phenotype screening program to obtain some indication of the function of the individual proteins [4]. Altogether these facts justify choosing *B. subtilis* as a model system for functional genomics, demonstrating the transformation of a genome sequence into the physiology of a living organism.

In 1975, O'Farrell [5] and Klose [6] introduced a new, highly sensitive protein separation technique that allowed the simultaneous study and separation of far more than 1000 proteins. This new technique relies on the typical features of proteins, their isoelectric point, and their molecular weight, which together bring each single protein to a unique position on a two-dimensional (2D) polyacrylamide gel. A few years later Fred Neidhardt and Ruth VanBogelen, the founders of the "physiological proteomics of bacteria", were the first authors to introduce this powerful approach into bacterial physiology ([7] for review).

We started our work with this new approach at the beginning of the 1980s, as we began to analyze the response of *B. subtilis* to stress and starvation, the environmental conditions that bacteria typically encounter in nature. We used the 2D separation technique to get a global view of the stress response, and found a dramatic change in the gene expression program of *B. subtilis* in response to stress or starvation [8–10]. This approach, recently adorned with the catchy term "proteomics" [11], has made remarkable progress during the last five years, stimulated particularly by the recent genome sequencing efforts. The sequencing of the *B. subtilis* genome provided a large amount of new information. Many predictions were derived from the genome sequence, including that of a large number of new regulators and regulons with still unknown functions, a great number of ABC transporters and secreted proteins [3]. However, these predictions based solely on the genome need experimental verification by the approaches of functional genomics. In this review article, a proteomic view of cell physiology of *B. subtilis* is presented as a model of how to transfer genome sequence information into the "real life" of *B. subtilis*.

2
A Proteome Map of *B. subtilis* as an Experimental Tool for Analyzing Cell Physiology

In order to use proteomics for analyzing cell physiology, a comprehensive 2D protein map depicting almost the entire proteome is the basic tool. First of all the genome sequence can be used to present a theoretical protein map derived solely from the genome sequence. Two main peaks were found: approximately two thirds of all proteins are located in the neutral or weak acidic region and the remainder in the alkaline region [12]. Many proteomic studies have ignored this alkaline fraction, focusing more on the pH range of 3–10 or even 4–7.

Furthermore, many proteins are secreted to fulfill their functions at the cell surface or extracellularly. This indicates one limitation of the method. In contrast to DNA arrays that may cover the entire genome on one single DNA chip, the depiction of the total proteome on one single gel is not possible; instead one has to rely on the combination of individual sub-proteomic fractions. When extracellular or cell-wall associated proteins are considered in addition to neutral and alkaline proteins, the majority of proteins synthesized in a bacterial cell can be visualized, and this technique may be used for physiological studies that focus on the cell as a whole [12, 13] (Fig. 1).

The alkaline fraction contains many proteins with more than two membrane-spanning domains [12]. However, no intrinsic membrane protein has been found among the 42 alkaline proteins identified so far [14]. This finding shows another limitation of the technique. Because a procedure for the complete visualization of all membrane-bound proteins is still lacking, 20–25% of all proteins are missing in proteomic studies. Our master gel contains almost 500 entries organized in a 2D protein database named Sub-2D [15] which is available via internet (http://microbio2.biologie.uni-greifswald.de:8880/sub2d.htm). In the following sections I shall show how these proteome data are used to gain new information on cell physiology.

3 From Proteome to Cell Physiology: Two Main Groups of Proteins

From a physiological point of view, two main groups of proteins can be distinguished: vegetative proteins synthesized particularly during growth, with mainly house-keeping functions, and proteins induced in response to stress and starvation, with mainly adaptive functions against stress or starvation. In depicting the entire proteome this physiological feature has to be taken into consideration. It will not be possible to visualize all proteins (no matter how sensitive the techniques) because a large portion of the genome is activated only when environmental stress or starvation stimuli or other extracellular signals are present.

Quantitative evaluation of the protein composition of growing cells revealed that biochemical functions are known for most of the abundant vegetative proteins. They have house-keeping functions typical of growing cells: in glycolysis, in the tricarboxylic acid cycle, in the synthesis of amino acids or nucleotides, in translation or in protein folding [12]. It is interesting to note that the transcription of these highly expressed genes correlates with the direction of replication of the *B. subtilis* chromosome. To consider functional groups of vegetative proteins more systematically, we can open a textbook of biochemistry and follow it chapter by chapter. This is done as an example for almost all proteins involved in amino acid biosynthesis [16] (Fig. 2) or in the basic routes of carbon catabolism such as glycolysis and TCA cycle (Fig. 3). The enzymes involved in amino-acid

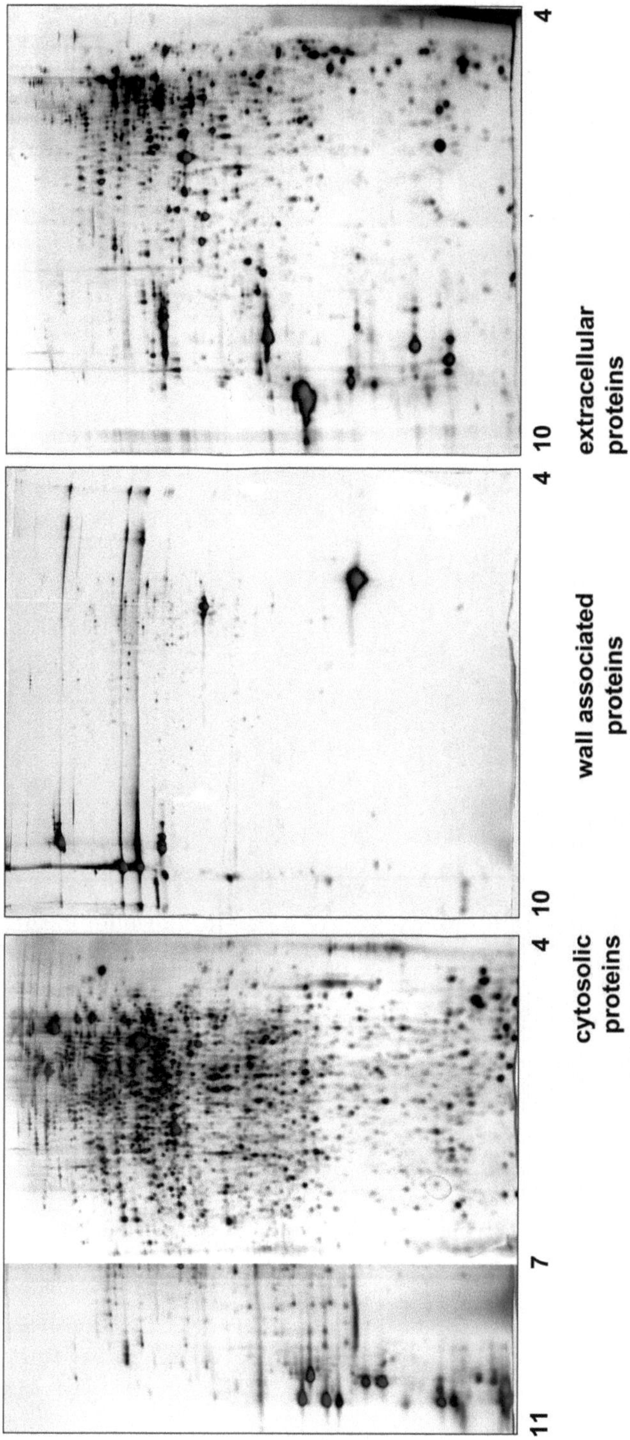

Fig. 1. Towards the total proteome of *B. subtilis* – cytosolic proteins (alkaline and acid/neutral fraction), cell-wall associated and extracellular proteins (see [12, 13])

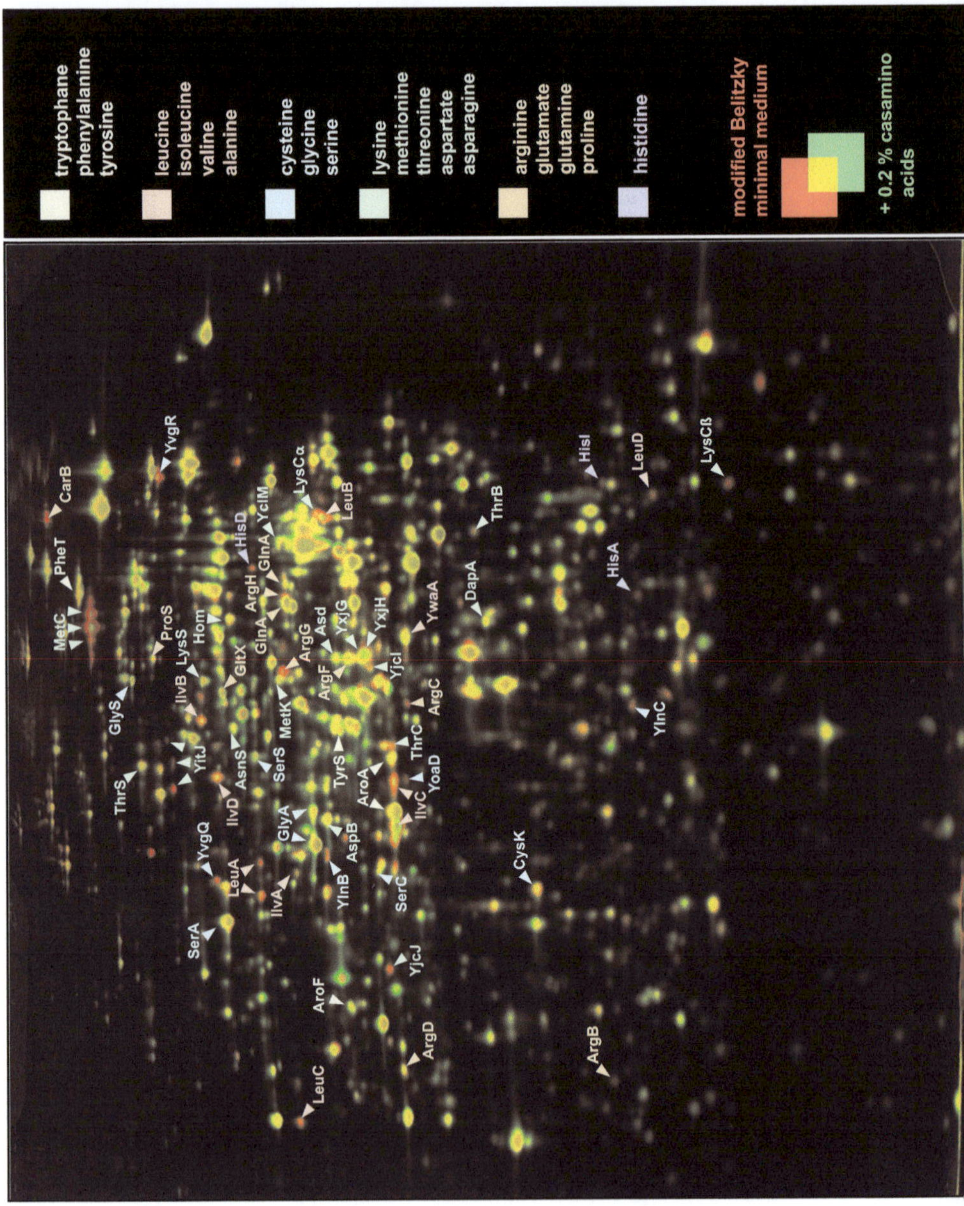

Fig. 2. Enzymes involved in amino acid biosynthesis in *B. subtilis* (from [16])

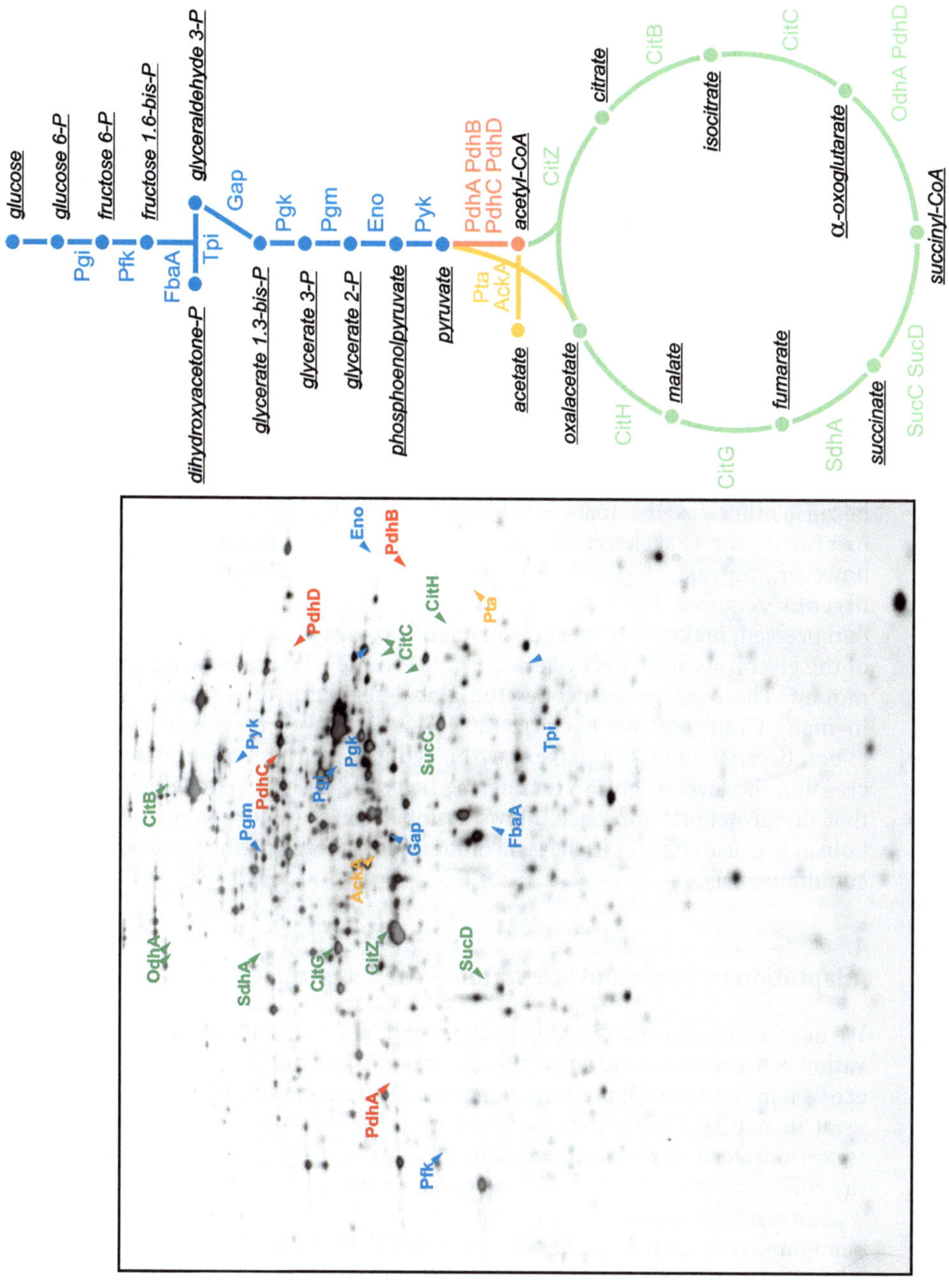

Fig. 3. Enzymes involved in glycolysis and the TCA cycle of *B. subtilis* (see [18]). *blue-labeled* – glycolysis; *red* – pyruvate dehydrogenase complex; *yellow* – overflow metabolism; *green* – TCA cycle

biosynthesis were found by comparing the proteomes of cells grown in the presence or absence of casamino acids. These functional subproteomic fractions allow a detailed study not only of some selected metabolic steps, but of entire metabolic pathways (as shown for the regulation of glycolysis/TCA cycle as an example).

Almost all enzymes involved in glycolysis and the TCA cycle were visualized on the 2D master gel at pH 4–7. This offers the chance to analyze simultaneously the regulation of glycolysis and the TCA cycle. We found that glycolysis is not a constitutive pathway, as is suggested in many textbooks. Cells were cultivated in the presence and in the absence of glucose. An excess of glucose stimulates glycolysis and strongly represses the TCA cycle (Fig. 4). The excess glycolytic intermediates cannot enter the TCA cycle because it is repressed, but instead they are metabolized via an "overflow" pathway and secreted into the extracellular medium as acetoine or acetate. Under the conditions of glucose excess ATP seems to be produced via substrate-level phosphorylation. This repression of the TCA cycle occurs only if glutamate is supplied simultaneously, because otherwise the anabolic functions of the TCA cycle would be necessary to provide the C-skeletons for amino acids [17–19]. In glucose-limited cells, however, the rate of glycolysis is strongly reduced, and the TCA cycle (now urgently required for the complete oxidation of glucose via acetyl CoA) is derepressed, making an overflow metabolism unnecessary. Finally, this control of the glycolysis and TCA cycle occurs only in the wild type and not in the *ccpA* mutant. The *ccpA* gene encodes the global regulator of catabolite repression in many Gram-positive bacteria [20]. This is, however, probably an indirect effect, because glucose uptake may be reduced in the *ccpA* mutant, thus decreasing the level of glycolytic intermediates [18]. This example demonstrates that the proteomic approach allows a global view of entire branches of metabolism because changes in the level of many enzymes/proteins can be considered simultaneously.

4 Adaptation to Stress and Starvation – The Adaptational Network

We have been more interested in proteins induced in response to stress and starvation because stress and starvation are the rule and not the exception in natural ecosystems and the adaptation to these environmental stimuli is crucial for survival in nature. Continuous fluctuations from slow-growing to non-growing states, long-term survival strategies in a non-growing state, or even cell death, are the consequences of bacterial life in harsh natural ecosystems [21].

As a result of this longstanding interplay between bacteria and a continuously changing set of environmental stimuli, a very complex adaptational network has been developed. This is a most characteristic feature of a bacterium, and its detailed understanding is crucial for understanding microbial physiology in general. At first, it is essential to look for those environmental stimuli that have limited *B. subtilis* growth in nature and thereby forced the complex adaptational strategies to maintain viability even during stress or starvation. Starvation for essential nutrients such as carbon/energy, nitrogen or phosphorus sources or oxy-

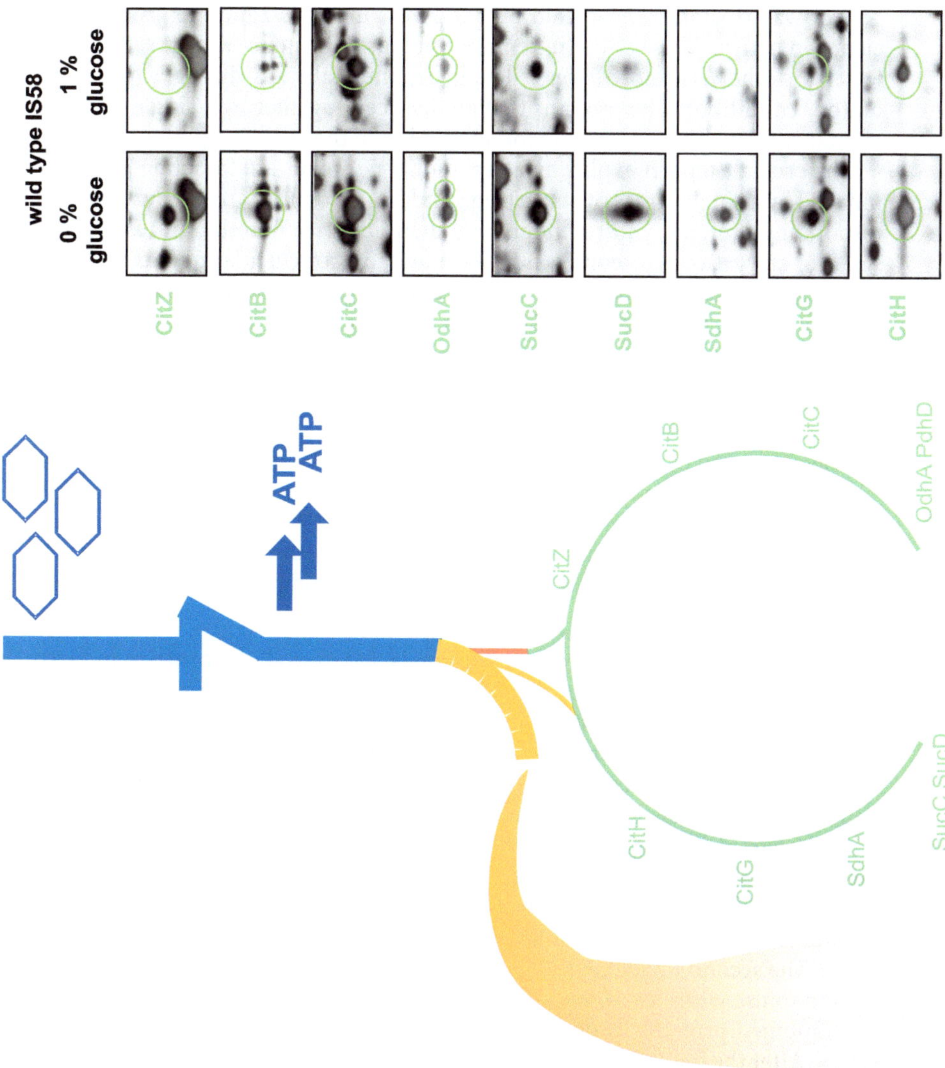

Fig. 4. A model of the regulation of glycolysis and the TCA cycle of *B. subtilis* during glucose excess. Glucose activates glycolysis and represses the TCA cycle (*shown on the right*). Therefore, excess glycolytic intermediates can not enter the TCA cycle, but have to be secreted into the extracellular medium as a result of an overflow metabolism (for details see text). ATP is mainly produced via substrate phosphorylation

gen, or water/osmotic, heat, cold or acid stress may contribute to growth deficiency in nature. These environmental stimuli usually induce a large set of genes which are otherwise more or less silent in exponentially growing cells under laboratory conditions. The proteins that are induced together by a single environmental stimulus are responsible for adaptation to the stimulus [22].

The crucial steps in exploring adaptational networks are to define the genes induced by a single stress or starvation stimulus, to identify and analyze the corresponding proteins, and to understand their adaptive function in response to stress or starvation. Proteomics is an excellent tool for elucidating this network, dissecting it into its individual components and studying the various adaptive functions of these components. The basic steps for exploring the modules of the entire network are:

1. The definition of stimulons (a stimulon is the entire set of proteins/genes induced or repressed by one stimulus) [23]
2. The dissection of stimulons into single regulons, the basic modules of global gene expression (a regulon consists of a set of genes distributed on the genome, but controlled by a global regulator)
3. The analysis of regions that overlap between different regulons (modulons), an essential step towards exploring complex adaptational networks (see Fig. 5)

The first step in analyzing stress adaptation is to define all the proteins induced by the stress or starvation stimulus, because these induced proteins are together responsible for adaptation to the stress. For defining stimulons the dual channel imaging technique [24] was developed to facilitate the search for proteins belonging to stimulons or regulons. This technique allows a rapid allocation of proteins to functional or regulation groups (stimulons, regulons) simply by looking for red-colored (newly synthesized) or green colored (repressed) proteins. Two digitized images of 2D gels have to be generated and combined in alternate additive color channels. The first one (densogram) showing accumulated proteins visualized by silver staining or some other staining techniques is false-colored green. The second image (autoradiograph) showing the proteins labeled during a 5-min pulse with ^{35}S-L-methionine is false-colored red. When the two images are combined, proteins accumulated and synthesized in growing cells are colored yellow. After the imposition of a stress/starvation stimulus, however, proteins not previously accumulated in the cell but newly induced are colored red. Identifying red proteins is a simple technique for finding all proteins induced by a single stimulus and thus defining the stimulon structure.

Heat stress, for instance, induces more than 100 proteins, some of which are of known function while others are of totally unknown function (Fig. 6). For these unknown proteins a preliminary prediction of their function may be feasible: Because of their classification as members of the heat stress stimulon they will somehow be involved in adaptation to heat stress. A similar technique can be used for a preliminary functional prediction of unknown proteins involved in adaptation to osmotic, acid, oxidative stress, etc. Proteins repressed by heat stress can be visualized by their green color, which indicates that they are present in the cell and probably still active, but no longer being synthesized (Fig. 6).

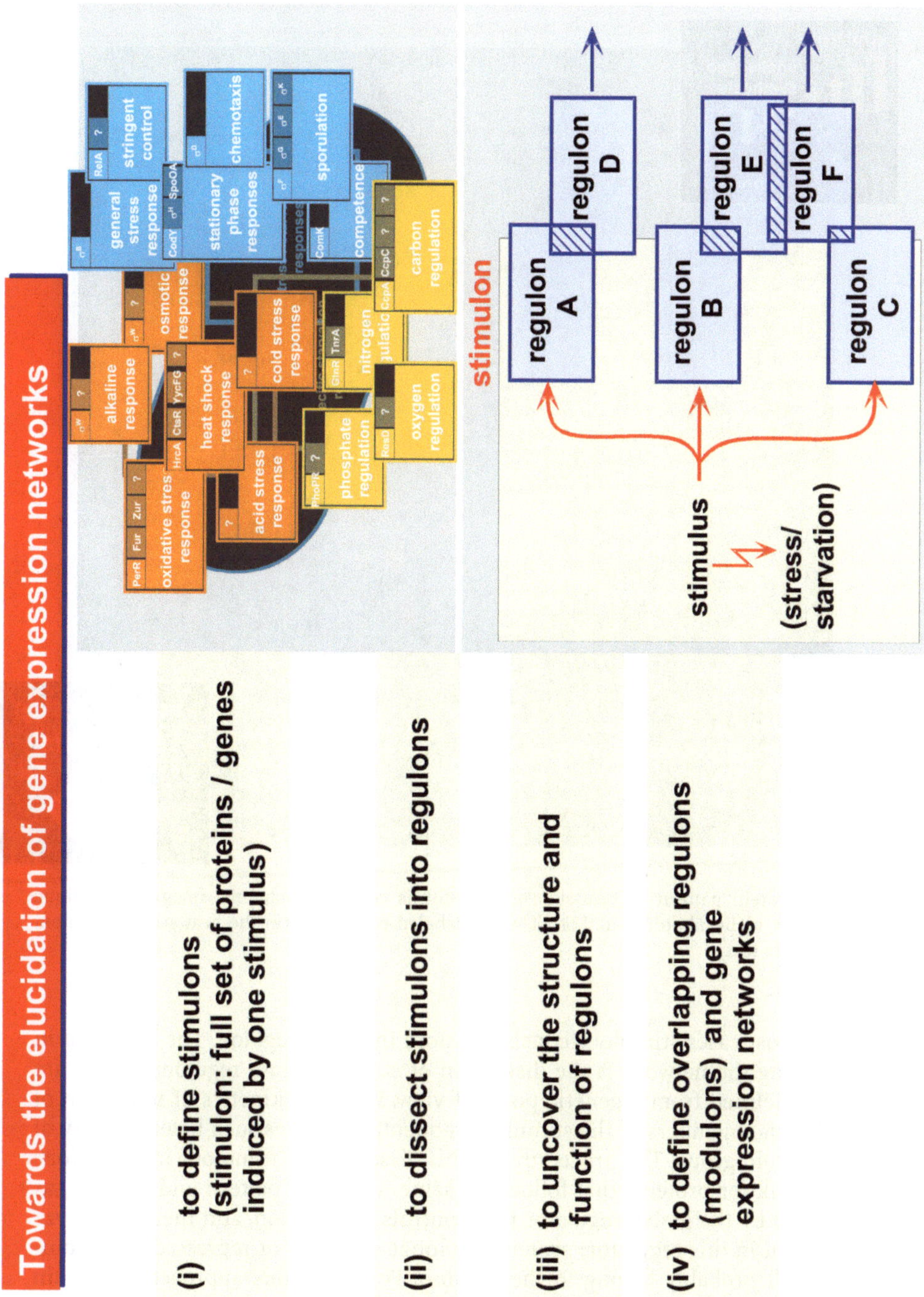

Fig. 5. Towards the elucidation of genes expression networks: from stimulons via regulons and modulons (overlapping regulons) to the network (see text)

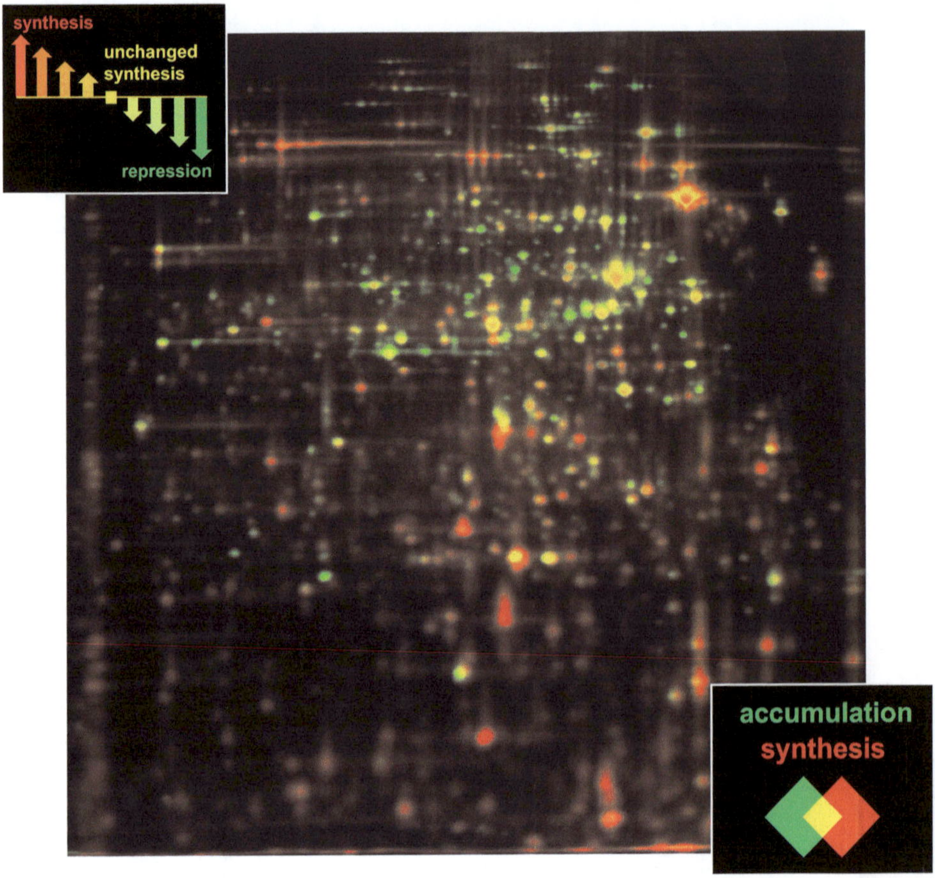

Fig. 6. Protein pattern of heat-shocked B. subtilis cells. Dual-channel imaging technique according to Bernhardt et al. [24]. The red-labeled proteins form the heat stress-stimulon. Details see text

In most cases stimulons consist of more than one regulon. The next step in exploring the network is the dissection of stimulons into regulons, which are better defined from a genetic point of view. Different kinetics of induction of individual members of the stimulon frequently indicates such heterogeneity of the protein group. The procedure for this dissection of stimulons into regulons is to look for proteins that follow the same induction pattern and kinetics, as dictated by the global regulator that controls the regulon, and then to analyze a mutant in this regulator; proteins no longer induced or repressed in the mutant will probably belong to the regulon. Following this approach two main regulon groups were found: proteins induced by only one stress or starvation stimulus with a specific adaptive function against that stimulus, and more general stress or starvation proteins induced by a different set of environmental stimuli.

The specific adaptive functions of the first group, the *stress-specific proteins* are:

- To neutralize the stress factor
- To adapt to its presence
- To repair damage caused by the stress stimulus

Starvation-specific proteins, on the other hand, may allow uptake of the limiting substrate with very high affinity, a search for alternative substrates not used in the presence of the preferred one, replacement of the limiting substrate by others, or moving to new nutrients by chemotaxis.

In addition to these stress- or starvation-specific proteins induced by a single stimulus, a totally different set of stress or starvation stimuli may induce the same set of proteins. This pattern of induction by stress or starvation indicates that such proteins may have a relatively non-specific, but nevertheless essential, protective function under stress, regardless of the specific growth-restricting signal. Therefore, these proteins have been called *general stress proteins* [8, 10, 25].

Most of the single stimulons within the network consist of specific and non-specific regulons such as the osmotic stress response, the heat- or acid-stress response or the response to glucose, oxygen, or phosphate starvation. One exception is oxidative stress, which does not induce a general stress response but only the specific oxidative-stress stimulon. This interplay between specific and general stress/starvation responses is shown for the phosphate starvation stimulon (Fig. 7). In addition to the σ^B-dependent general stress response (see below) phosphate starvation induces the main phosphate starvation-specific regulon, which is controlled by a two-component system with PhoP as the sensor kinase and PhoR as the response regulator [26]. These proteins are no longer induced by phosphate starvation in a *phoPR* mutant [27]. Identification of these proteins allows insights into the function of this starvation-specific regulon. It turns out that the proteins strongly and specifically induced by phosphate starvation have specific functions that help the cell to adapt to phosphate starvation. They are involved in a high-affinity phosphate uptake system when the external phosphate concentration is limiting (e.g., Pst-system), in the utilization of alternate phosphate sources (e.g., extracellular phosphatase PhoB and D), or in the replacement of the phosphate in structural components (the phosphate-containing teichoic acid is replaced by the phosphate-lacking teichuroic acid, function of TuaD). A similar proteomic approach can be used for exploring the adaptive function of the specific responses to glucose, nitrogen or oxygen starvation if mutants in global regulator genes are available (see [28, 29]).

A different kind of adaptation is achieved by the *specific stress responses*. We can use heat stress as an example. In addition to the general stress proteins, also induced by phosphate starvation, heat shock induces a small group of heat-specific stress proteins, the chaperones (GroEL-machine and DnaK machine) that are characteristic of heat stress and controlled by the global HrcA repressor molecule [30]. In an *hrcA* mutant the proteins are not repressed but are synthesized at a high level even at normal temperatures. The most essential specific protective function is to assist protein folding under heat stress or refolding of partially heat-denatured proteins. A second group of heat-specific stress proteins are the

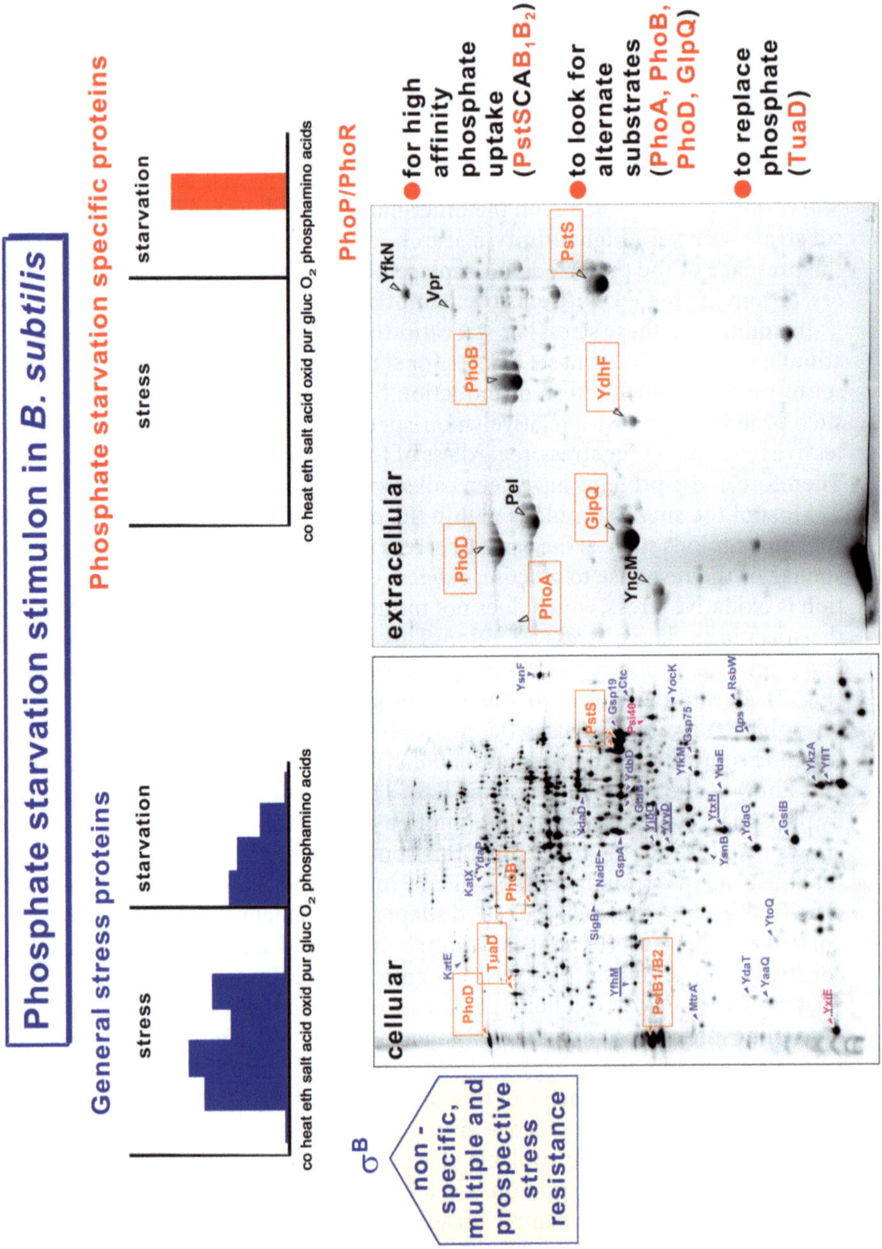

Fig. 7. The phosphate starvation stimulon of *B. subtilis* consists of general stress proteins also induced by phosphate starvation (σ^B-regulon, σ^B-dependent proteins *marked in blue*) as well as of phosphate starvation specific proteins (PhoPR-Regulon *marked in red*). The stress/starvation induction profile of both groups and the function of the proteins are indicated. co – control, heat – heat stress, eth – ethanol stress, salt – NaCl stress, acid – pH 5.5, oxid – oxidative stress (H_2O_2), pur – puromycin, starvation for gluc – glucose, O_2 – oxygen, phospH – phosphate, amino acids. Details see text

Clp proteins (ATPases, proteases), controlled by the global CtsR repressor [30]. Only three genes/operons, *clpP*, *clpE*, and the *clpC* operon, form this small regulon. All Clp proteins were identified as heat-inducible on 2D gels. In addition to their chaperone function ClpC, and probably also ClpE, in combination with the proteolytic component ClpP, form the Clp proteases, which degrade unstable or hopelessly denatured proteins in order to remove toxic protein waste (Fig. 8).

The most striking protein group within the heat stress or phosphate-starvation stimulon, however, which is strongly induced by heat but also by many other stress and starvation stimuli, is controlled by the alternative sigma factor σ^B [31, 32]. When we started analyzing this regulon 15 years ago we did not know

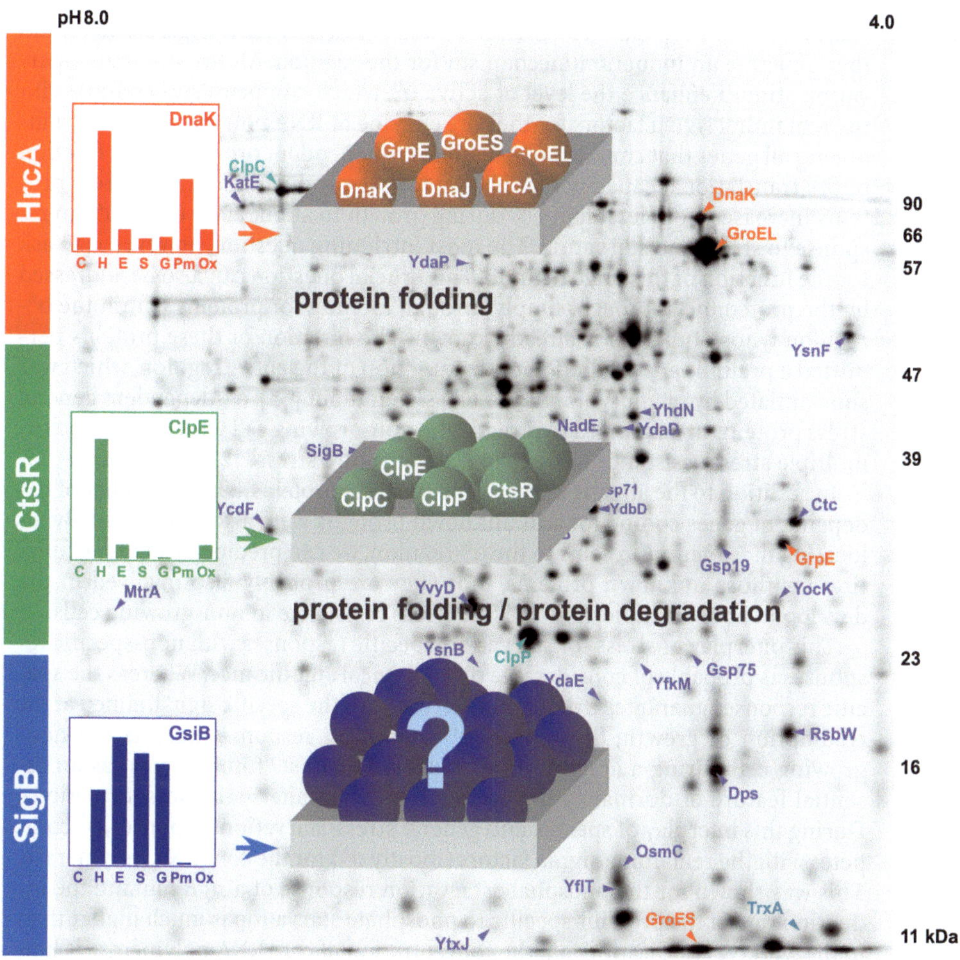

Fig. 8. The heat stress stimulon of *B. subtilis* (details see text). Induction profile: C – control, H – heat, E – ethanol, S – salt, G – glucose starvation, Pm – puromycin, Ox – oxidative stress (H_2O_2). The function of the HrcA- and CtsR regulon is known, but what about the σ^B-dependent general stress response?

anything about the genes of the regulon, the global regulator, or the signal transduction pathway – or indeed the physiological role of the regulon. We only noticed that many proteins on the 2D protein gels were strongly induced by different stress and starvation stimuli. A systematic use of the proteomic approach allowed us to uncover part of the structure/size and function of an unknown regulon, in this case the σ^B regulon. The general procedure is as follows. Proteins that follow the stress and starvation induction pattern typical of σ^B are good candidates for regulon members. About 50 proteins were found that followed this typical induction profile and which were no longer stress- or starvation-inducible in a *sigB* mutant (Fig. 9). To obtain information on the stress induction mechanism, it was necessary to look for the target sequence of this regulon. All σ^B-dependent genes so far identified by the proteomic approach are preceded by σ^B-dependent promoters with typical –10 (GGGTAT) and –35 (GTTTAA) regions. We could then imagine an induction mechanism for the regulon. All the stress and starvation stimuli enhance the level of active σ^B, which competes successfully with the remaining sigma factors for the core enzyme of RNA polymerase, thereby inducing all genes that contain an accessible σ^B-dependent promoter. The fraction of the translational capacity that is used for the production of these stress proteins increases from less than 1% during growth to about 20 or even 30% in response to stress and starvation. The most intriguing question, however, is to ask for the function of this huge regulon. This crucial question can also be addressed by the proteomic approach, simply through looking for proteins within the σ^B-regulon whose function is already known. The function of these proteins permitted a preliminary prediction of the function of the entire regulon, which was substantiated by phenotypic studies of a *sigB* mutant [33]. σ^B-dependent general stress proteins are expected to provide the non-growing cell with a non-specific, multiple stress resistance in "anticipation of future stress" [22, 25, 32, 34].

In addition to the already known general stress proteins, a large number of σ^B-dependent genes coding for still unknown proteins were also discovered. By allocating these genes/proteins to the σ^B-regulon, we can predict an adaptive function for those unknown proteins, too. They are probably also involved in the development of multiple, non-specific stress resistance in non-growing cells.

This interplay of stress- or starvation-specific responses with non-specific responses is probably of considerable physiological significance. Whereas the specific responses guarantee a direct interaction with the specific signal, aimed at the resumption of growth, the general σ^B-dependent response protects the non-growing cell during a lengthy survival period against "future stress" as an essential feature of dormancy in vegetative cells (an alternative to sporulation). During this interplay of specific and general stress/starvation responses, σ^B competes with the remaining sigma factors (mostly σ^A) for the core RNA polymerase. This was shown for the phosphate starvation response of a *sigB* mutant: the induction rate of the proteins specific to phosphate starvation is much higher than in the wild type, indicating that the concentration of core RNA polymerase is limiting in the cell [27, 35].

In this section the σ^B regulon of *B. subtilis* has been presented as a model to show the value of the proteomic approach for defining the size and physiological function of an unknown regulon. This is a question of current importance, be-

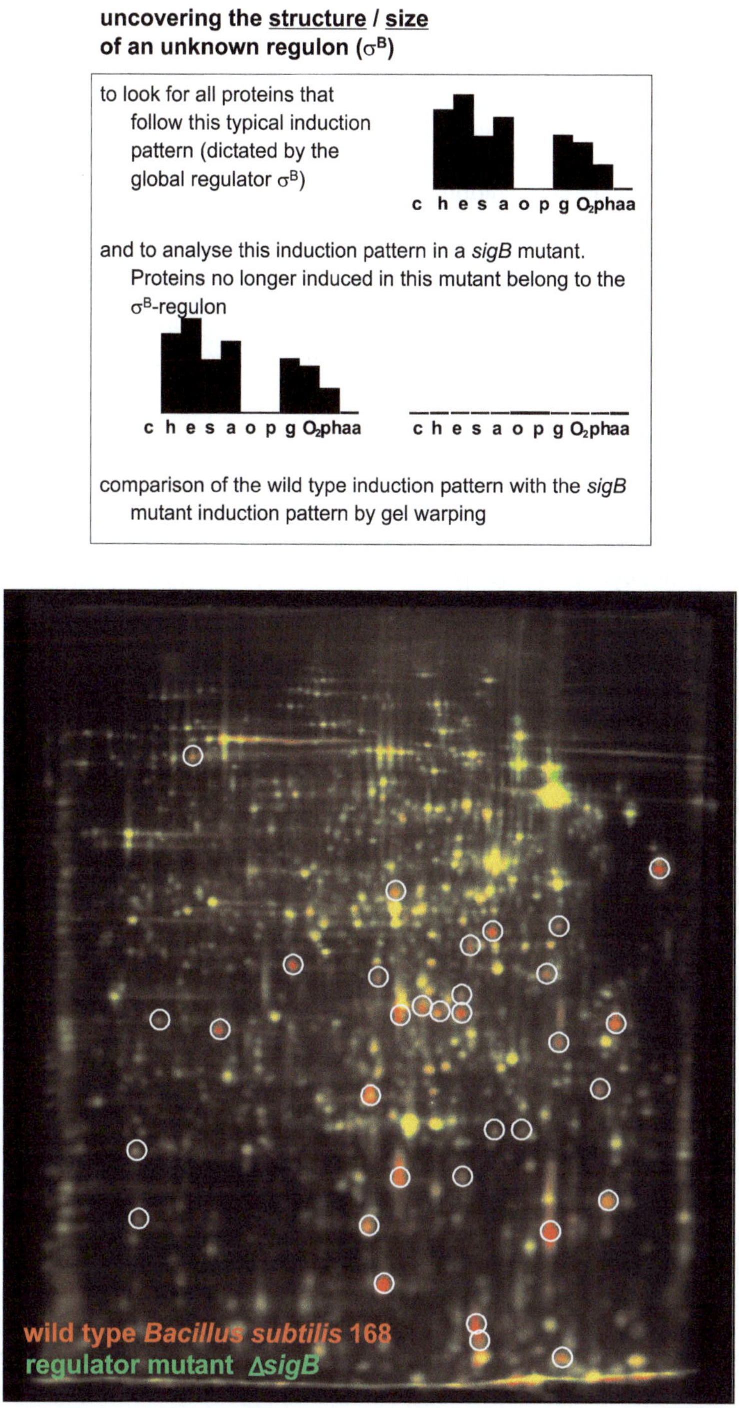

A

B

Fig. 9. A Strategies to find σ^B-dependent proteins. B Use of the Decodon software package delta 2D to define the σ^B-dependent stress proteins induced by heat stress in the wild type only (red), but not in the sigB mutant (green). The proteins marked in red may belong to the σ^B-regulon cause many previously unknown regulators (probably global) have been discovered by genome sequencing, with each controlling its own, still unidentified, regulon [3]. New alternative sigma factors, unknown two-component regulatory systems, and additional global activators or repressors and the regulons they control await detailed characterization. The approach described here for the σ^B-regulon can also be used for analyzing the size and function of these remaining unidentified regulons provided that the physiological conditions that activate the regulon are known. The steps are as follows:

1. To construct a mutation in the regulatory gene
2. To allocate the genes to the regulon by proteomics (or more recently by DNA array techniques)
3. To predict the function of the regulon from the function of already known proteins belonging to the regulon
4. To confirm this prediction by phenotypic studies of mutants
5. To predict the function of the still unknown proteins of the regulon that might be expected within the context of the physiological role of the regulon to which the genes belong (see [34]).

To dissect the entire genome into its basic modules of global gene regulation, and to put the still unknown proteins into these functional categories, is a good approach for an overall, albeit still preliminary, prediction of the function of most of the unidentified proteins (Fig. 10). This procedure, however, is not sufficient for understanding the global regulation of gene expression in the adaptational network, because the single regulons do not exist independently but, instead, are tightly connected, forming the adaptational network. Genes that are controlled by two global regulators function as overlapping elements between the single regulons that connect the regulons together. Proteomics again is a good strategy for finding such connecting elements between regulons, which is shown for our model regulon.

A search for proteins that follow the stress and starvation induction pattern typical of σ^B regulation but with some modifications provides candidates for such a search. For example, the reason for the atypical induction of the σ^B-dependent ClpC protein by puromycin or oxidative stress is the dual control by σ^B and CtsR that connects the σ^B-regulon with the CtsR-regulon. Similarly, the reason for the atypical induction of the σ^B-dependent general stress protein YvyD by amino acid starvation is the dual control of *yvyD* by σ^B and σ^H, which forms a link between the σ^B-regulon and the σ^H-regulon [22].

The tight integration of the single regulons into a group of related regulons may have a physiological significance. In principle, all σ^B-dependent genes have to follow the σ^B-dependent stress/starvation induction pattern. However, a fine-tuning of the expression of some genes by environmental stimuli not related to σ^B can occur if additional regulatory elements interacting with other global or specific regulators have been acquired. By this fine-tuning of gene expression each gene can be expressed at a level required for an optimum response to the environmental stimuli.

These results show that a group of *stress-specific regulons*, of *starvation-specific regulons* and of *general stress or/and starvation regulons* form the complex

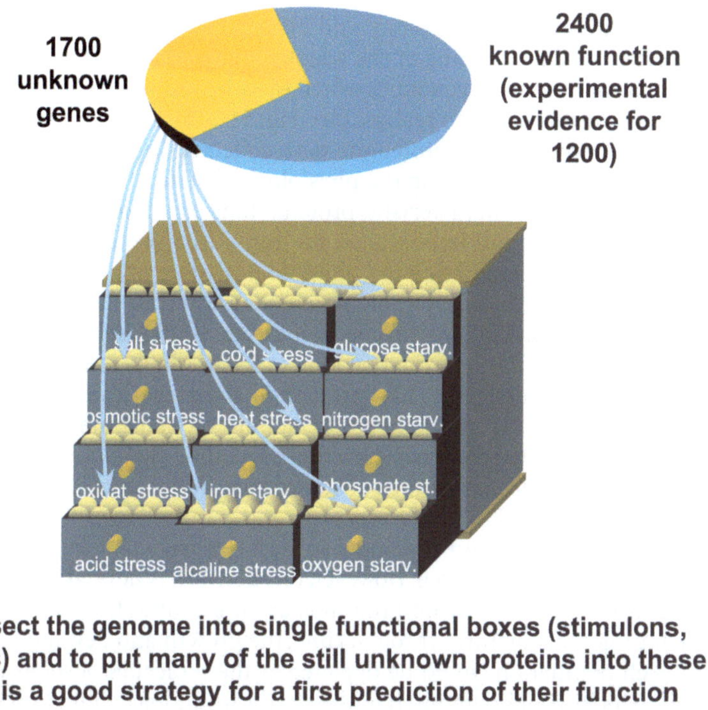

Fig. 10. The allocation of still unknown proteins to functional groups (regulons, stimulons) is a good strategy for a first prediction of the function of those unknown proteins

network. As already shown for the σ^B model case, the network itself seems to be stabilized by genes that connect two or more regulons. Following the route from regulon to regulon is a good method for analyzing this network. From the σ^B-regulon there is a direct path to the σ^H-regulon, which in turn is connected with regulons involved in sporulation [36] and also with the RelA-regulon [22]. By this "regulon walking" the complex adaptational network can be explored step by step.

To summarize this point, different methods can be used to explore the network. First, all the single stress- and starvation-specific, or more general, regulons have to be studied. In the second step, all these basic modules have to be integrated into the network, and finally the overlapping areas will provide the cement that stabilizes the entire network. This systematic approach allows us to discern all functional boxes, the stimulons regulated by the most essential growth-restricting stimuli, their dissection into regulons, and finally their overlapping areas in the network. The final aim might be the presentation of a "virtual cell", including "clickable" information on the single stimulons and regulons, their genes and proteins, their DNA target sequence, their induction/repression ratios by environmental stimuli at the protein or mRNA level (see later), their protein stability, their integration into the network by means of overlapping elements, and finally the functions of all proteins arranged in the boxes in response to stress or starvation stimuli. Such a cell model connecting DNA sequence information with cell physiology and biochemistry would provide a most comprehensive understanding of cell adaptation to stress or starvation, an excellent example of functional genomics which would show how to proceed from genome sequence information to real life (see Fig. 5).

The complete set of proteomic data collected in this adaptational network can be used for diagnosing the physiological state of the cell. Most stimulons/regulons contain characteristic proteins that can provide information on the physiological stimuli imposed on the cell. This "proteomic signature" [37] is a useful tool for predicting the physiological state of a cell population, e.g., of cells grown in a bioreactor, or of cells harvested in natural ecosystems such as biofilms. Marker proteins will indicate if the cells suffered from heat (GroEL, DnaK, etc.) or oxidative stress (KatA, AhpC, etc.) or other stresses or starvation for nutrients. This "proteomic signature" can also be used to predict the molecular mechanism of action of some unknown antibacterial drugs. Drug-treated cells that show the "proteomic signature" for oxidative stress, for instance, indicate that the substance has produced oxidative damage of cell structures.

The network we have presented so far has given a picture of a single moment in the life of a *B. subtilis* cell. However, this network is not static but dynamic, with sequential gene expression programs as an essential feature. If one assembled such proteomic pictures through time, the growth and development of a bacterial cell population would be depicted at the molecular level as in a "movie of life", which would demonstrate the synthesis and accumulation of each single protein. This is shown for growing cells entering a glucose-starvation-induced stationary phase, a typical environmental situation for *B. subtilis* in its natural environment (Fig. 11). Again, the dual channel imaging technique [24] is an excellent tool for visualizing changes in the synthesis and level of each protein separated on the 2D gel. During exponential growth most of the proteins are colored yellow because synthesis (red color) and accumulation (green color) are in steady state. After glucose exhaustion, however, most of the vegetative proteins typical of growing cells are no longer synthesized, and their color then changes from yellow to green. At the same time a great number of red-colored proteins are induced in a sequential order, representing the glucose-starvation stimulon that consists of starvation-specific and general stress responses [38]. With a basic knowledge of bacterial physiology one can reach a comprehensive understanding of what is happening in the cell. Among the reactions specific to glucose-starvation, one can find a diminution in glycolysis, an induction of genes for the usage of alternative carbon sources such as acetoine, and the induction of gluconeogenesis (Fig. 12). Among the more general starvation responses one can follow the kinetics of the negative and positive stringent response, the σ^B-dependent general stress response, and some other reactions (CodY-dependent, Spo0A-dependent), and finally proteins that indicate that some cells have already started the sporulation program. However, there are also unknown responses to be explored by future studies, because the function of many proteins is still unidentified. In many cases, however, the allocation of some of these unknown proteins to already known regulons may provide a provisional indication of their functions in glucose-starved cells.

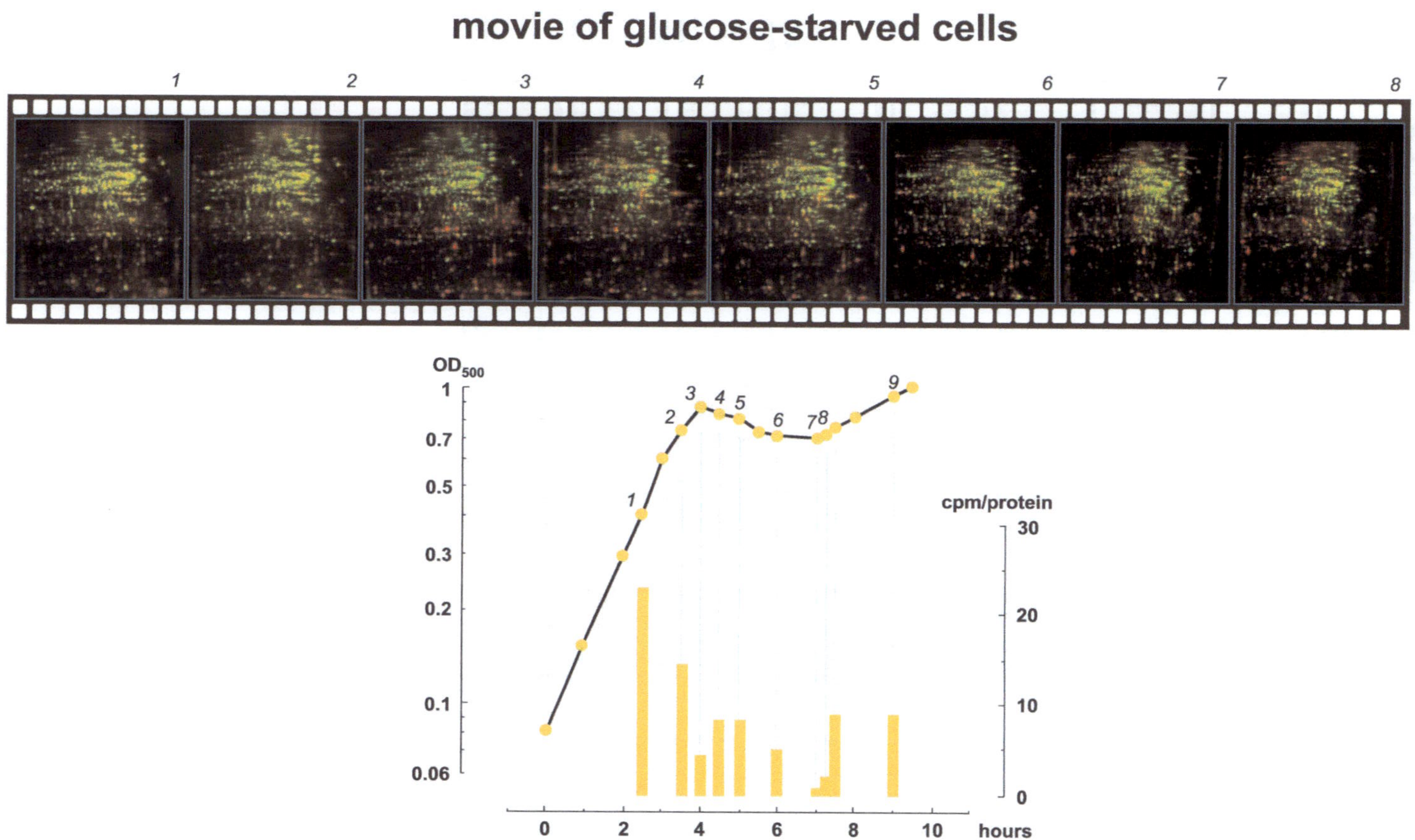

Fig. 11. Dynamics of the protein synthesis profiles in growing and glucose-starved cells – "movie" of growth and glucose starvation (see [38]). Growth curve (o. D.) and ^{35}S-*L*-methionine incorporation (10^6 cpm/60 μg protein)

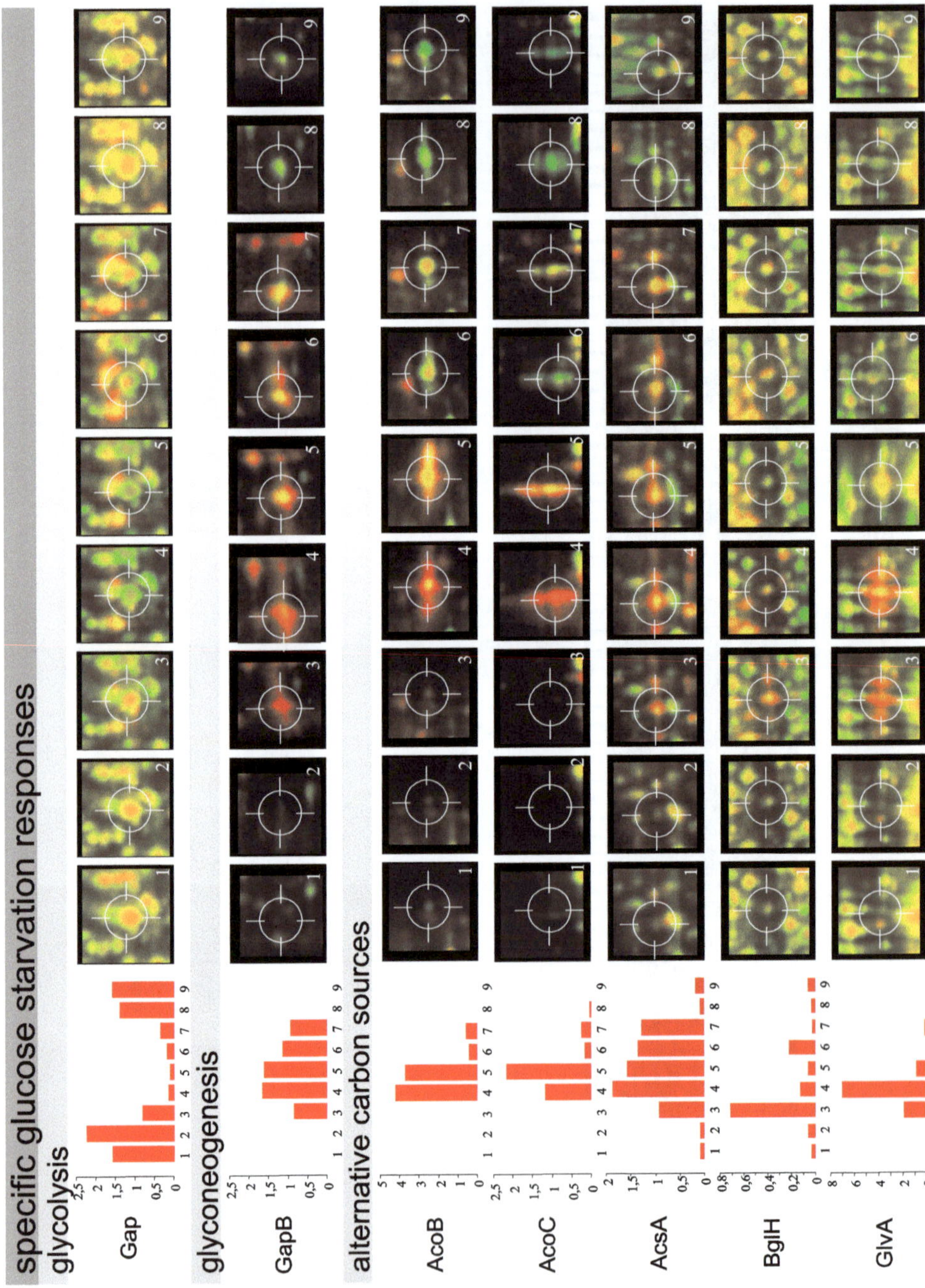

Fig. 12. Continued (for caption see p 80)

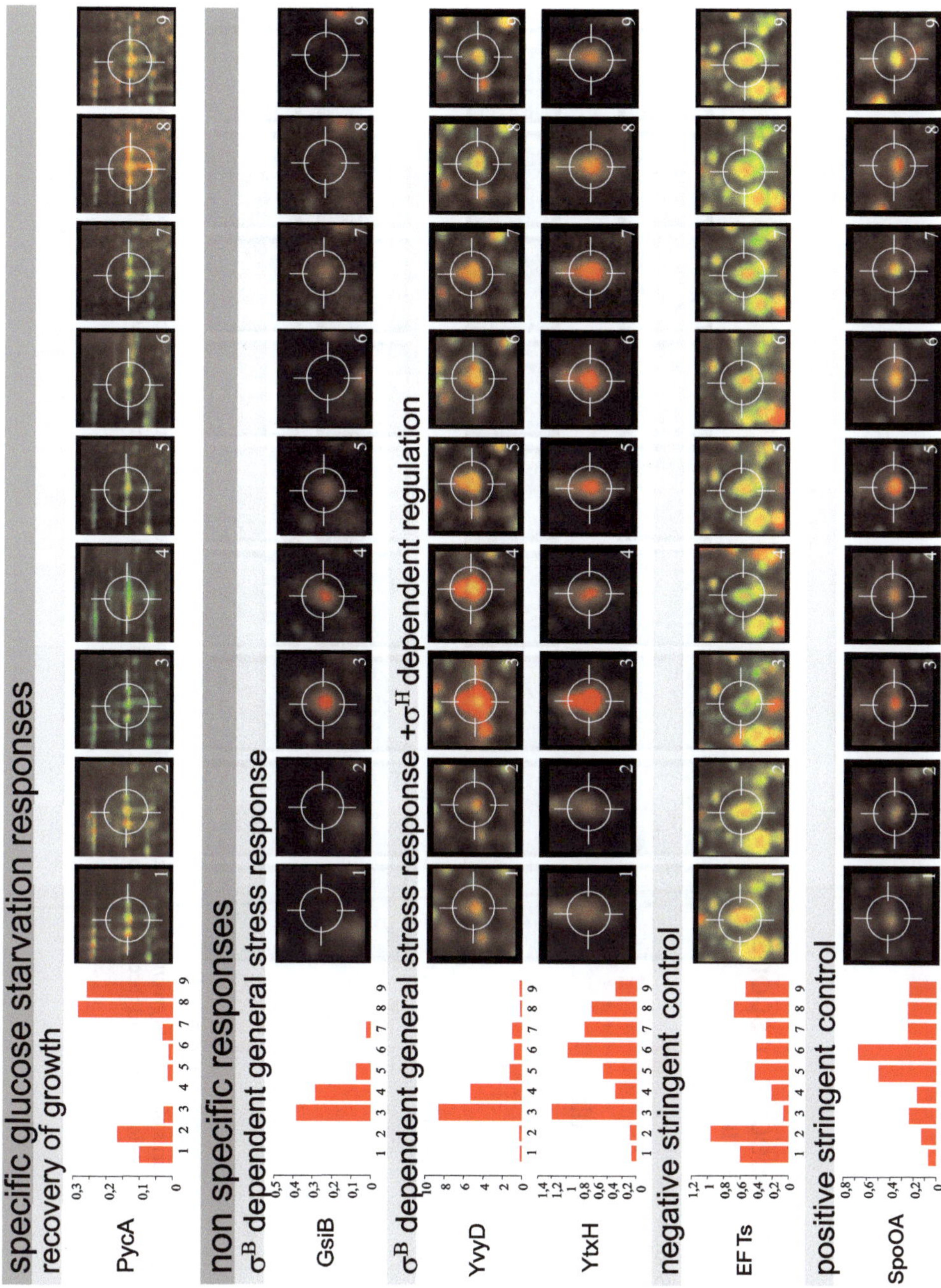

Fig. 12. Continued (for caption see p 80)

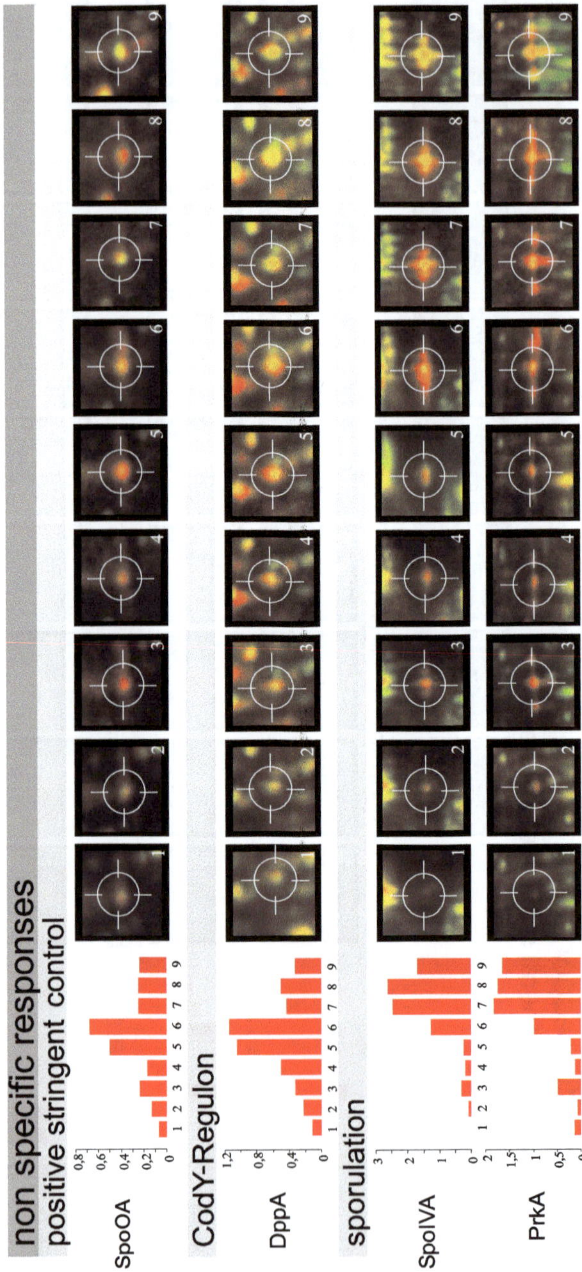

Fig. 12. Patterns of amount (*green*) and synthesis (*red*) of general stress and glucose-starvation specific responses during growth (1+2), glucose starvation (3–7) and recovery of growth (8+9). *Left*: protein synthesis rates of single proteins (from [38])

Finally, using the "three color code" offered by the dual channel imaging technique [24] it is possible not only to follow the induction/repression kinetics of entire regulation groups but also to study the "fate" of each single protein with the three colors: "red" – newly induced but not yet accumulated; yellow – synthesized as well as accumulated; and green – no longer synthesized but still present and probably active in the cell. This is shown for AcoB – one subunit of the acetoine-dehydrogenase – which is necessary for the utilization of acetoine as a secondary carbon source when glucose is exhausted (Fig. 13).

In conclusion, the transition of *B. subtilis* from growing cells to glucose starvation is accompanied by an almost complete reorganization of gene expression. Almost 400 proteins change their color from yellow to green because their synthesis has been switched off in the stationary phase. This behavior is typical of vegetative proteins synthesized during growth. Because the growing cell entering the stationary phase of growth probably contains an excess of these vegetative proteins, their continued synthesis in the non-growing state would be wasteful of resources during energy- and carbon-source limitation. On the other hand,

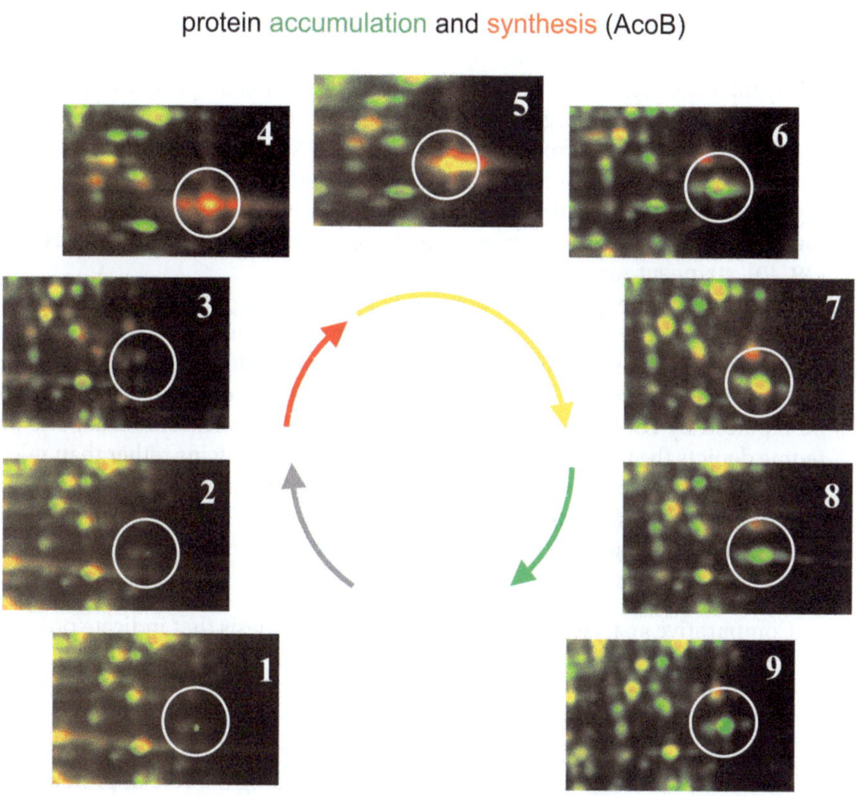

Fig. 13. Kinetics of synthesis (*red*) and accumulation (*green*) of acetoine dehydrogenase subunit AcoB during growth (1, 2), glucose starvation (3–7) and recovery of growth (8, 9)

more than 150 proteins with different protective functions are induced in a sequential manner in the course of glucose-starvation-specific and more general responses.

The conclusion that the gene expression network is totally reprogrammed has been confirmed and extended by the application of DNA macro-array nylon membranes that contain PCR-products of each single gene. These DNA arrays allow genome-wide mRNA profiling. Almost all genes transcribed under these special circumstances are involved, either repressed or induced/derepressed during glucose starvation [38]. In contrast with the proteome approach, which depicts only a subpopulation of the proteins, the expression of all genes can be followed. Both approaches have specific advantages (see below), but the combination of the two techniques offers exciting possibilities for a much expanded understanding of cell physiology.

5 Limitations and New Perspectives of the Proteomic Approach

A comparison of the proteomic approach with DNA array techniques showed that for the allocation of genes to regulons, or for the study of genome-wide gene expression, the DNA array technique records changes in gene expression more completely than the proteome approach. Too many proteins of the cell – particularly intrinsic membrane proteins but also low abundance proteins, very acid or alkaline proteins, very large or small proteins – cannot be visualized by the conventional proteomic approach. By contrast, the DNA array technique allows a reliable depiction of all bacterial genes expressed at a definite time point, including information on putative operon structure etc. Only around half of the genes found to be active by DNA arrays were detected by proteomics (e.g., σ^B-regulon [39, 40], stringent response [41]).

Accordingly, the proteomic approach is now being replaced by transcriptomics, where genes belonging to regulons or stimulons, or genome-wide gene expression patterns, are analyzed [29, 42–44]. Nevertheless, proteomic studies will maintain their central importance when they focus on aspects which cannot be replaced by DNA arrays or which over-extend transcriptome data. The proteome depicts the final level of gene expression, since proteins rather than mRNA molecules are the players of life. mRNA molecules are highly unstable intermediates on the path from the genes to their destination. The fields of proteomics that cannot be replaced by transcriptomics or which extend or complement transcriptional data are:

- Comparative analyses of mRNA and protein synthesis that indicate posttranscriptional or translational regulation, when the mRNA level does not necessarily reflect the rate of protein synthesis
- Analysis of protein sorting/targeting
- Analysis of posttranslational modification of proteins
- Analysis of protein turnover and stability, including the role of proteases
- Analysis of proteins in non-growing cells that are still present but no longer synthesized ("green" proteins)

Protein quality control, including protein folding, stability, and turnover, and also posttranslational modification, repair of proteins, or protein degradation, are becoming more and more prominent, and can be followed on the proteomic scale or at the level of single proteins.

In the final section, the capacity of proteomics to explore protein secretion at the proteomic level has been selected to show one of the new perspectives of proteomics.

6 Proteomics, Cell Architecture and Protein Sorting/Targeting on a Proteomic Scale

Genome-wide mRNA profiling only provides information on the transcription of the genome, but provides no information about the activity, level, or final destination of the gene products, the proteins. Proteomics, on the other hand, provides information about the concentration of most of the individual proteins in the cell. A big challenge for future proteomic studies is to follow the route and fate of all proteins from the polysomes to their final destination. However, again knowledge of all the proteins, their concentration in the cell, and their functions is not enough to make an organism viable.

Only in rare cases do proteins act independently of other proteins. Cellular life is characterized by a complex protein interaction network, with many proteins taking part in highly ordered protein complexes that need a particular spatial structure for their function. Recent efforts have been aimed at detecting these protein-protein interactions during proteomic studies, e.g., by using the yeast two hybrid system, affinity chromatography, and other techniques [45, 46]. Metabolism does not function in a bag of enzymes, but instead seems to be highly organized in complex structures that allow the channeling of intermediates from enzyme to enzyme [47]. An understanding of this ordered interplay of individual proteins is crucial to an understanding of what constitutes life.

The majority of proteins act in the cytoplasm, but many proteins have specific sorting signals that lead them to their final destination outside the cytoplasm, e.g., to the cytoplasmic membrane, the membrane cell wall interspace, the cell wall, or even the extracellular space. Two main approaches to visualize these final destinations may be considered. One is to look for localization of the proteins by sophisticated microscopic techniques after labeling the proteins with fluorescent dyes. In practice, some proteins may have a fixed position in the cell, while others move, e.g., from cell pole to cell pole as shown by time lapse microscopy [48–50]. A "molecular topology" of the cell may be the intriguing outcome of these novel studies. An excellent example of how this technique can be used on a proteomic scale is given in this issue (Chap. 8). The more "classical" proteomic approach analyses the protein composition of subcellular fractions, the membrane fraction, the periplasmic fraction, and the "outer membrane" proteins for Gram-negative bacteria, the cell wall-associated proteins or the extracellular proteins. As mentioned earlier, proteomics visualization of the membrane-bound proteins is an unsolved problem which remains a challenge for future research.

B. subtilis, a soil-living bacterium, seems to be optimized for protein secretion, probably because extracellular polymeric substances are typical nutrients for this species. van Dijl et al. [51] inspected the entire genome sequence of *B. subtilis* for specific secretion signals and predicted four distinct pathways for protein export from the cytoplasm. It is probable that around 300 proteins are exported, and the majority follow the major Sec pathway for protein secretion. Only a few proteins appear to be transported via the more specialized pseudopilin export pathway for competence development, or via a pathway relying on ABC (*A*TP-*b*inding-*c*assette) transporters, or via the recently identified twin-arginine translocation (Tat) pathway that allows the transport even of folded proteins [51]. Again, the proteomic approach is an excellent means of providing experimental evidence for genome-based signal peptide predictions. In a recent paper entitled "A proteomic view on genome-based signal peptide predictions", Antelmann et al. [52] recognized almost 200 extracellular proteins by 2D gel electrophoresis. Of these, 82 were identified (see also [53]). These included 41 proteins with a signal peptide and a predicted type I signal peptidase cleavage site. Surprisingly, the remaining 41 extracellular proteins were predicted to be cell-associated [52], either because of the absence of a signal peptide (22 proteins) or because of the presence of specific cell retention signals in addition to a signal peptidase cleavage site (19 proteins) (see [52]).

A collaboration between a laboratory well known for its expertise in protein secretion and a proteomic laboratory afforded the opportunity of gaining new information on protein secretion on a proteomic scale. By the use of a wide collection of mutants and the cultivation of cells under different growth and starvation conditions, new results on the mechanism and regulation of protein secretion can be obtained, as shown by a few selected examples (Fig. 14).

6.1 Mechanisms of Protein Secretion

- The PhoD precursor, an extracellular phosphatase, was predicted to contain a twin arginine motif in its N-terminal signal peptide (van Dijl, private communication). Experimental evidence for this prediction was provided by the analysis of a mutant lacking the *tatCd* gene encoded downstream of the *phoD* gene. Whereas the amount of the remaining extracellular proteins was unaffected in the extracellular proteome of the *tatCd* mutant, secretion of PhoD was completely prevented [54] (Fig. 15).
- Pre-lipoproteins are characterized by a specific "lipobox" containing a conserved cysteine residue in the signal peptide which is the target for lipid modification by the lipoprotein diacylglyceryl transferase (Lgt) and subsequent processing by the lipoprotein-specific type SPase II (LspA). After processing the diacylglyceryl residue is anchored into the cytoplasmic membrane [55, 56]. Inspection of the genome sequence revealed 114 putative pre-lipoproteins [51] containing the characteristic lipobox. Surprisingly, eight potential lipoproteins were found in the extracellular space. All of them lacked the cysteine residue at the *N*-terminus, indicating a proteolytic shaving after their processing by SPase II or an alternative protease [52]. As expected, lipoprotein shedding was

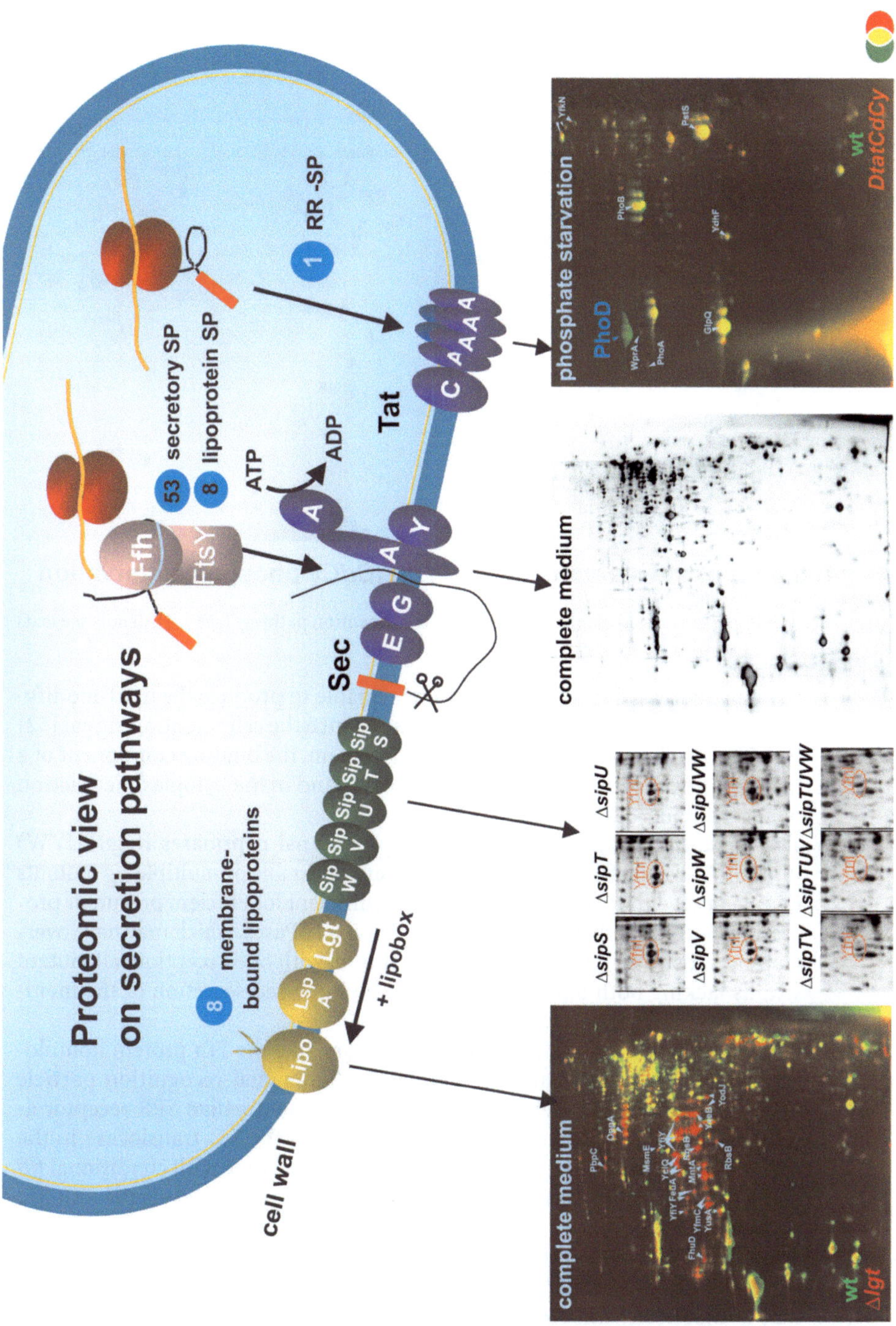

Fig. 14. Combined efforts of a genetic and proteomic group to elucidate the mechanisms of protein secretion at a proteomic scale (for details see text, according to [51, 52, 54, 58])

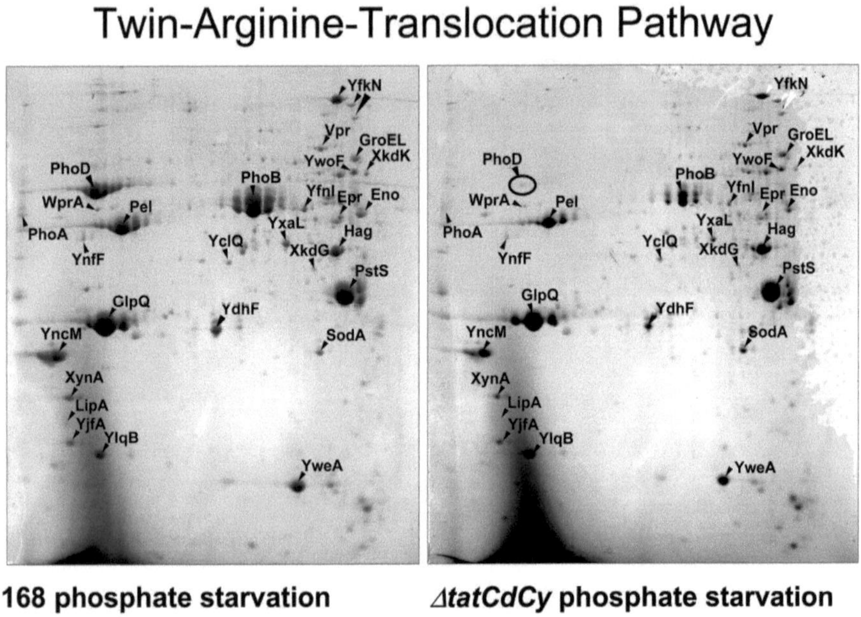

Fig. 15. PhoD is secreted via the twin arginine translocation pathway (see [54], details see text)

strongly increased in a *lgt* mutant that is unable to produce the lipid-modifying enzyme diacylglyceryl transferase and hence the cell retention signal [52] (see Fig. 14). The non-modified Pre-PstS protein, the binding component of a high affinity phosphate uptake system, was found in the cytoplasmic fraction of a *lgt* mutant [52].

- The *B. subtilis* genome encodes five type I signal peptidases (SigSTUVW) [57, 58]. A systematic proteomic study of single, double or multiple *sig* mutants revealed that either SipS or SipT is almost sufficient for efficient precursor processing. Specific "substrates" for the single type I SPases, which may have overlapping substrate specificities, were not found, with one exception: all mutant strains lacking both SipT and SipV showed impaired secretion of the membrane protein YfnI [52] (see Fig. 14).
- Cytoplasmic chaperones and targeting factors, like the Ffh protein homologous to the 54-kDa subunit of the mammalian signal recognition particle (SRP) and the FtsY protein homologous to the mammalian SPR receptor a-subunit, facilitate targeting of the pre-proteins to the Sec-translocase in the membrane [59, 60]. Studies of the extracellular proteome with conditional *ffh* and *secA* mutants revealed that many proteins are translocated via the SRP-Sec pathway [53].
- The chaperone and peptidyl prolyl *cis-trans* isomerase PrsA seems to be essential for the proper folding of extracellular proteins [61], because a *prsA* mutant is drastically impaired in protein secretion [62]. Presumably any misfolded proteins that are secreted are degraded by cell-surface located proteases. In the absence of SPase II the secretion of non-lipoproteins was

reduced, indicating that an essential lipoprotein may be required for proper protein folding of extracellular proteins. Because PrsA is a lipoprotein this chaperone, which is probably no longer fully active in the pre-protein conformation, is a good candidate for such a predicted role.

6.2
Physiology and Regulation

A systematic study of the pattern and level of extracellular proteins during growth in L-broth medium revealed a marked increase in protein secretion when cells entered the stationary phase of growth. This is not surprising because one of the main functions of these extracellular proteins seems to be to provide alternate growth substrates when the preferred substrates are exhausted. This is the physiological role of many extracellular hydrolases such as α-amylase, various β-glucanases, chitosanase, pectate lyases, xylanases, various proteases, nucleases, or lipases. The kind of nutrient limitation has a crucial influence on the extracellular protein pattern which differs, for instance, totally in different stationary-phase cells, depending on whether they are grown in L-broth medium, or are glucose or phosphate-starved. The mechanisms that keep the genes silent in growing cells are partly known. In glucose-starved cells, genes encoding extracellular enzymes such as α-amylase or β-gluconases are derepressed because the global repressor CcpA is no longer active, but in phosphate-starved cells a phosphorylated and thereby activated form of the response regulator PhoP is required for gene activation [26]. Furthermore, a two-component regulatory system with DegS as a sensor kinase and DegU as a response regulator seems to be involved in the regulation of enzymes in cells grown to stationary phase in L-broth medium or in response to carbon source/energy source starvation. Strong overproduction of many extracellular proteins was found in a *degU32(hy)* [52] mutant that is characterized by a hyperactive (hyperphosphorylated) response regulator DegU [63, 64]. The extracellular proteins overproduced in this *degU32(hy)* mutant strain are probably under direct DegU control. The different protein pattern of glucose- or phosphate-starved cells is not surprising because the proteins/enzymes produced in the two cell types have a totally different physiological function, either to provide new carbon/energy sources when glucose is exhausted or to provide new phosphate sources when phosphate is limiting [27]. This explains why only a subset of extracellular proteins is dependent on the source of nutrient limitation, illustrating that proteomics should always be considered in close relationship to cell physiology. The physiology of the cell does not allow one to cover the total extracellular protein fraction no matter how sensitive and sophisticated the separation/identification techniques are because many genes remain silent under specific physiological circumstances.

Accordingly, the large number of proteins that is presumed to be secreted by the SecA or Tat pathway but not yet detected by the proteomic approach is rather surprising. A more systematic study of the low-abundance extracellular proteins, or a more systematic study of starvation conditions (e.g., oxygen, nitrogen) will probably provide more members of the extracellular proteome. However, is the provision of new and alternate substrates really the only role of extracellular proteins in *B. subtilis*, or are there other functions that have not yet been considered? Another function of protein secretion is to organize cell-cell communication. In Gram-positive pathogens such as *Staphylococcus aureus* or *S. epidermidis*, an important function of cell-surface located or extracellular proteins is to adhere to host cell surfaces, to invade host cells or to act as virulence factors (see [65]). It may be that environmental conditions exist in the natural ecosystems of *B. subtilis* that require specific extracellular proteins, but are not taken into account in artificial laboratory studies. Examples of such a "natural lifestyle" of *B. subtilis* that are not replicated in laboratory conditions would be growth in the neighborhood of plant roots (rhizosphere), growth in the form of microcolonies on soil particles, or even biofilms [66, 67]. It may be that: (i) some extracellular proteins such as exopolysaccharide-forming enzymes are required for the formation of biofilms and (ii) the pattern of extracellular proteins in such biofilms is different from that found in a growing cell population in liquid medium. A proteomic approach, complemented by a DNA array approach, is needed to address these problems, which are crucial from a physiological point of view.

A final physiological item considered in this chapter is secretion stress. The experimental overexpression of heterologous genes whose products will be secreted may overload the secretion capacity of the cell. In this case the cell activates a two-component system (CssR/S) whose gene products deal with this secretion stress [68]. Two stress-inducible chaperones/proteases, HtrA and YvtA, belong to this regulon, which may help to overcome protein secretion stress by refolding denatured proteins or by degrading hopelessly destroyed (and hence toxic) proteins. A proteomic approach has demonstrated that HtrA, a chaperone and stress-inducible protease, is the major protein of the regulon found in the extracellular space, because it probably has lost its membrane-anchoring domain [69].

These examples should demonstrate that proteomics is not only a part of protein biochemistry and protein analysis. Technical developments, such as novel high throughput techniques based on mass spectrometry, are extremely important and form the basis for many fields of proteomics application. The combination of proteomics with cell physiology or molecular genetics, however, (which we call "physiological proteomics") opens new perspectives for functional genomics to reach a comprehensive understanding of life processes at a molecular level.

7 Outlook

Proteomics offers new opportunities to observe events in the cell never seen before by looking at cells in a new and wider context. Not just a few interesting proteins, but almost all the proteins of the cell, are the subject of those studies. However, one can get lost in the mass of data that have to be considered. The pitfall may be that one remains at the surface of the problem, and does not leave the descriptive level – which is, however, necessary for a comprehensive understanding of the phenomena to be studied. That was in our mind when we started to analyze the response of *B. subtilis* cells to stress and starvation more than 15 years ago. Visual inspection of the 2D protein pattern in response to stress and starvation revealed obvious changes that were not visible by means of other tech-

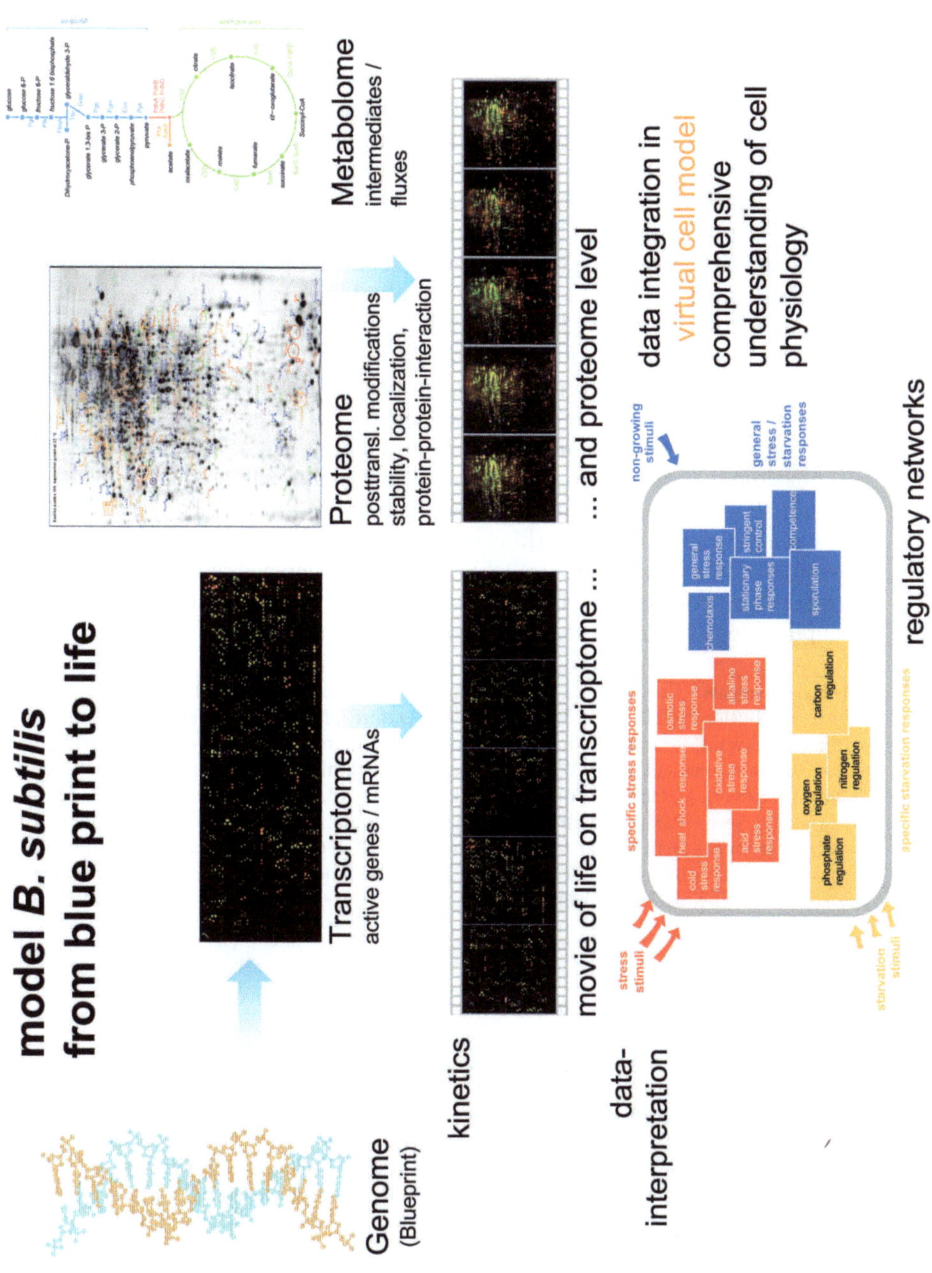

Fig. 16. From the genome sequence to cell physiology

niques available at that time, indicating that a huge, and probably essential, stress response had been activated. This finding initiated a systematic study of the role of what is probably one of the most comprehensive stress and stationary phase responses of Gram-positive bacteria, the σ^B response (see [22, 32] for a review). This example from our own work should illustrate the fact that when proteomics is used to pinpoint the most interesting problems, and is supplemented by more detailed experiments using biochemical or genetical approaches, the intriguing potential of proteomics can be fully exploited.

This review chapter should show that the sequencing of entire genomes yields only a first, introductory chapter in the "book of life". From genome sequencing a long route must be followed to "bring this genome sequence to life" (see Fig. 16). The route passes through genome-wide mRNA profiling, followed by "physiological proteomics" which provides information on protein function, protein distribution inside and outside the cell, protein-protein interactions, degradation, repair, or posttranslational modification, and it aims at a new understanding of cell physiology in its entirety. To assemble and analyze all the data collected by functional genomics, bioinformatics is urgently required, which has the ambitious aim of simulating the working of cells by mathematical models. Within the ensemble of functional genomics, which reach from genome sequencing to bioinformatics, proteomics will retain its central position because it deals – as no other discipline does – directly with the players of life, the proteins.

Acknowledgements. The author is very grateful to his co-workers and many PhD students for their contribution to this work, Michael Yudkin (Oxford), Lindsay Winkler and Haike Antelmann (Greifswald) for critical comments, and Jörg Bernhardt for his support in preparing the figures. I thank particularly U. Völker (Marburg), R. Schmid and K. Altendorf (Osnabrück), E. Ron (Tel Aviv), J. Hacker (Würzburg), S. Bron and J. M. van Dijl (Groningen), M. Sarvas (Helsinki), S. Seror (Paris), and others for a fruitful cooperation. This work has been supported by grants from the Deutsche Forschungsgemeinschaft (DFG), the European Union, the Bundesministerium für Bildung und Forschung (BMBF), the Land Mecklenburg-Vorpommern, and the Fonds der Chemischen Industrie to M. H.

8
References

1. Fleischmann RD, Adams MD, White O, Clayton RA, Kirkness EF, Kerlavage AR, Bult CJ, Tomb JF, Dougherty BA, Merrick JM et al. (1995) Science 269:496
2. Sonenshein AL, Hoch JA, Losick R (eds) (2002) *Bacillus subtilis* and its closest relatives: from genes to cells. ASM Press, Washington DC
3. Kunst F, Ogasawara N, Moszer I, Albertini AM, Alloni G, Azevedo V, Bertero MG, Bessieres P, Bolotin A, Borchert S, Borriss R, Boursier L, Brans A, Braun M, Brignell SC, Bron S, Brouillet S, Bruschi CV, Caldwell B, Capuano V, Carter NM, Choi SK, Codani JJ, Connerton IF, Danchin A et al. (1997) Nature 390:249
4. Schumann W, Ehrlich SD, Ogasawara N (2001) (eds) Functional analysis of bacterial genes: a practical manual. Wiley, Weinheim
5. O'Farrell PH (1975) J Biol Chem 250:4007
6. Klose J (1975) Humangenetik 26:231
7. Neidhardt FC, VanBogelen RA (2000) Proteomic analysis of bacterial stress responses. In: Storz G, Hengge-Aronis R (eds) Bacterial stress responses. ASM Press, Washington DC, p 445

8. Richter A, Hecker M (1986) FEMS Microbiol Lett 36:69
9. Hecker M, Heim C, Völker U, Wölfel L (1988) Arch Microbiol 150:564
10. Hecker M, Völker U (1990) FEMS Microbiol Ecol 74:197
11. Wasinger VC, Cordwell SJ, Cerpa-Poljak A, Yan JX, Gooley AA, Wilkins MR, Duncan MW, Harris R, Williams KL, Humphery-Smith I (1995) Electrophoresis 16:1090
12. Büttner K, Bernhardt J, Scharf C, Schmid R, Mäder U, Eymann C, Antelmann H, Völker A, Völker U, Hecker M (2001) Electrophoresis 22:2908
13. Antelmann H, Yamamoto H, Sekiguchi J, Hecker M (2002) Proteomics (in press)
14. Ohlmeier S, Scharf C, Hecker M (2000) Electrophoresis 21:3701
15. Bernhardt J, Büttner K, Coppée JY, Lelong C, Ogasawara N, Scharf C, Vagner V, Schmid R, Völker U, Hecker M (2001) The contribution of the EC consortium to the two-dimensional protein index of *Bacillus subtilis*. In: Schumann W, Ehrlich SD, Ogasawara N (eds) Functional analysis of bacterial genes: a practical manual. Wiley, Weinheim, p 63
16. Mäder U, Homuth G, Scharf C, Büttner K, Bode R, Hecker M (2002) J Bacteriol 184:4288
17. Fouet A, Sonenshein AL (1990) J Bacteriol 172:835
18. Tobisch S, Zühlke D, Bernhardt J, Stülke J, Hecker M (1999) J Bacteriol 181:6996
19. Ludwig H, Homuth G, Schmalisch M, Dyka FM, Hecker M, Stülke J (2001) Mol Microbiol 41:409
20. Stülke J, Hillen W (2000) Annu Rev Microbiol 54:849
21. Storz G, Hengge-Aronis R (eds) (2000) Bacterial stress responses. ASM Press, Washington DC
22. Hecker M, Völker U (2001) Adv Microb Physiol 44:35
23. Gottesman S, Neidhardt FC (1983) Global control systems. In: Beckwith J (ed) Gene function in prokaryotes. Cold Spring Harbor Laboratory, Cold Spring Harbor NY, p 163
24. Bernhardt J, Büttner K, Scharf C, Hecker M (1999) Electrophoresis 20:2225
25. Hecker M, Völker U (1998) Mol Microbiol 29:1129
26. Hulett FM (1996) Mol Microbiol 19:933
27. Antelmann H, Scharf C, Hecker M (2000) J Bacteriol 182:4478
28. Wendrich TM, Marahiel MA (1997) Cloning and characterization of a *relA/spot* homologue from *Bacillus subtilis*. Mol Microbiol 26:65
29. Yoshida K-I, Kobayashi K, Miwa Y, Kang CM, Matsunaga M, Yamaguchi H, Tojo S, Yamamoto M, Nishi R, Ogasawara N, Nakayama T, Fujita Y (2001) Nucl Acids Res 29:683
30. Schumann W, Hecker M, Msadek T (2002) Regulation and function of heat-inducible genes in *Bacillus subtilis*. In: Sonenshein AL, Hoch JA, Losick R (eds) *Bacillus subtilis* and its closest relatives: from genes to cells. ASM Press, Washington DC, p 359
31. Haldenwang WG (1995) Microbiol Rev 59:1
32. Price CW (2002) General stress response. In: Sonenshein AL, Hoch JA, Losick R (eds) *Bacillus subtilis* and its closest relatives: from genes to cells. ASM Press, Washington DC, p 369
33. Völker U, Maul B, Hecker M (1999) J Bacteriol 181:3942
34. Hecker M, Engelmann S (2000) Int J Med Microbiol 290:123
35. Pragai Z, Harwood CR (2002) Microbiology (in press)
36. Sonenshein AL (2000) Bacterial sporulation: a response to environmental signals. In: Storz G, Hengge-Aronis R (2000) (eds) Bacterial stress responses. ASM Press, Washington DC, p 199
37. VanBogelen RA, Schiller EE, Thomas JD, Neidhardt FC (1999) Electrophoresis 20:2149
38. Bernhardt J, Weibezahn J, Scharf C, Hecker M (2003) Genome Res 13:224
39. Petersohn A, Brigulla M, Hass S, Hoheisel J, Völker U, Hecker M (2001) J Bacteriol 183:5617
40. Price CW, Fawcett P, Ceremonie H, Su N, Murphy CK, Youngman P (2001) Mol Microbiol 41:757
41. Eymann C, Homuth G, Scharf C, Hecker M (2002) J Bacteriol 184:2500
42. Kobayashi K, Ogura M, Yamaguchi H, Yoshida K, Ogasawara N, Tanaka T, Fujita Y (2001) J Bacteriol 183:7365
43. Helman JD, Wu MFW, Kobel PA, Gamo FJ, Wilson M, Morshedi MM, Navre M, Paddon C (2001) J Bacteriol 183:7318
44. Fawcett P, Eichenberger P, Losick R, Youngman P (2000) Proc Natl Acad Sci USA 5:8063

45. Rain JC, Selig L, De Reuse H, Battaglia V, Reverdy C, Simon S, Lenzen G, Petel F, Wojcik J, Schachter V, Chemama Y, Labigne A, Legrain P (2001) Nature 409:211
46. Gavin AC, Bösche M, Krause R, Grandi P, Marzioch M, Bauer A, Schultz J, Rick JM, Michon AM, Cruciat CM, Remor M, Höfert C, Schelder M, Brajenovic M, Riffner H, Merino A, Klein K, Hudak M, Dickson D, Rudi T, Gnau V, Bauch A, Bastuck S, Huhse B, Leutwein C, Heurtier MA, Copley RR, Edelmann A, Querfurth E, Rybin V, Drewes G, Raida M, Bouwmeester T, Bork P, Seraphin B, Kuster B, Neubauer G, Superti-Furga G (2002) Nature 415:141
47. Mathews CK (1975) J Bacteriol 175:6377
48. Raskin DM, De Boer PA (1999) J Bacteriol 181:6419
49. Losick R, Shapiro L (1999) J Bacteriol 181:4143
50. Rudner DZ, Losick R (2001) Dev Cell 1:733
51. Tjalsma H, Bolhuis A, Jongbloed JDH, Bron S, van Dijl JM (2000) Microbiol Mol Biol Rev 64:515
52. Antelmann H, Tjalsma H, Voigt B, Ohlmeier S, Bron S, van Dijl JM, Hecker M (2001) Genome Res 11:1484
53. Hirose I, Sano K, Shioda I, Kumano M, Nakamura K, Yamane K (2000) Microbiology 146:65
54. Jongbloed JDH, Martin U, Antelmann H, Hecker M, Tjalsma H, Venema G, Bron S, van Dijl JM, Müller J (2000) J Biol Chem 275:41,350
55. Tjalsma H, Kontinen VP, Pragai Z, Wu H, Meima R, Venema G, Bron S, Sarvas M, van Dijl JM (1999) J Biol Chem 274:1698
56. Lesela S, Wahlstrom E, Kontinen VP, Sarvas M (1999) Mol Microbiol 31:1075
57. Tjalsma H, Noback MA, Bron S, Venema G, Yamane K, van Dijl JM (1997) J Biol Chem 272:25,983
58. Tjalsma H, Bolhuis A, van Roosmalen ML, Wiegert T, Schumann W, Broekhuizen CP, Quax WJ, Venema G, Bron S, van Dijl JM (1998) Genes Dev 12:2318
59. Honda K, Nakamura K, Nishiguchi M, Yamane K (1993) J Bacteriol 175:4885
60. Ogura A, Kakeshita H, Honda K, Takamatsu H, Nakamura K, Yamane K (1995) DNA Res 2:95
61. Kontinen VP, Saravs M (1993) Mol Microbiol 8:727
62. Sarvas M, Kontinen VP, Antelmann H, unpublished
63. Henner DJ, Yang M, Ferrari E (1988) J Bacteriol 170:5102
64. Kunst F, Debarbouille M, Msadek T, Young M, Mauel C, Karamata D, Klier A, Rapoport G, Dedonder R (1988) J Bacteriol 170:5093
65. Ziebandt AK, Weber H, Rudolph J, Schmid R, Höper D, Engelmann S, Hecker M (2001) Proteomics 1:480
66. Branda SS, Gonzales-Pastor JE, Ben-Yehuda S, Losick R, Kolter R (2001) Proc Natl Acad Sci USA 98:11,621
67. Hamon MA, Lazazzera BA (2001) Mol Microbiol 42:1199
68. Hyyrylainen HL, Bolhuis A, Darmon E, Muukkonen L, Koski P, Vitikainen M, Sarvas M, Pragai Z, Bron S, van Dijl JM, Kontinen VP (2001) Mol Microbiol 41:1159
69. Antelmann H, Darmon E, Noone D, Veening JW, Bron S, Kuipers OP, Devine KM, Hecker M, van Dijl JM (2003) Mol Microbiol (in press)

Received: April 2002

Adv Biochem Engin/Biotechnol (2003) 83: 93–115
DOI 10.1007/b11118

Proteomics of Bacterial Pathogens

Phillip Cash

Department of Medical Microbiology, University of Aberdeen, Foresterhill, Aberdeen AB32 6QX, Scotland. *E-mail: p.cash@abdn.ac.uk*

The rapid growth of proteomics that has been built upon the available bacterial genome sequences has opened provided new approaches to the analysis of bacterial functional genomics. In the study of pathogenic bacteria the combined technologies of genomics, proteomics and bioinformatics has provided valuable tools for the study of complex phenomena determined by the action of multiple gene sets. The review considers some of the recent developments in the establishment of proteomic databases as well as attempts to define pathogenic determinants at the level of the proteome for some of the major human pathogens. Proteomics can also provide practical applications through the identification of immunogenic proteins that may be potential vaccine targets as well as in extending our understanding of antibiotic action. There is little doubt that proteomics has provided us with new and valuable information on bacterial pathogens and will continue to be an important source of information in the coming years.

Keywords. Proteome, Bacteria, Protein, Pathogenesis, Antibiotic resistance, Antigens

List of Abbreviations

2DGE 2-Dimensional gel electrophoresis

1
Introduction

The technologies that support the rapidly developing field of proteomics have been extensively used in microbiology (see the following reviews for examples [1–3]). The extensive gene sequence data that are now available for many bacteria make bacterial proteomes eminently suitable for characterisation. The continuing reports of complete genome sequences for a variety of bacteria have fuelled the rapid developments in the study of microbial proteomics. Sixty-three complete microbial genomes are presently listed at the NCBI web site (www.ncbi.nlm.nih.gov/PMGifs/Genomes/bact.html) and a further 137 genomes are being actively sequenced and annotated.

In the field of proteomics there has been rapid progress in the development of new techniques to investigate the proteome. Novel high throughput methods based on mass spectrometry [4] and specialised protein arrays [5, 6] are being developed. These methods will complement or, in some instances, replace the traditional methods of classical proteomics, namely 2DGE for protein separation and peptide mass mapping for protein identification. Nevertheless the use of the latter methods still play a key role in defining bacterial proteomes [1] and have been used in many of the studies to be discussed below. Comprehensive descriptions of proteome and genomes are being established for a number of different bacteria. For example, with the small bacterium *Spiroplasma melliferum*, which is predicted to encode up to 1000 proteins, 456 of the proteins were resolved from unfractionated cell lysates using 2DGE and 98 of these proteins were assigned an identity [7, 8]. This review will consider the investigation of the proteomes of pathogenic bacteria. One of the key areas of interest to the medical microbiologist concerns how bacteria interact with their host and most significantly how they cause overt disease. The control of these processes in bacteria is liable to be polygenic and modern techniques of functional genomics, for example transcriptomics and proteomics, are ideal for the large scale screening of gene expression on a genome wide scale.

2
Bacterial Proteome Databases

Global analyses of bacterial proteomes have been a feature of a number of investigations. Ever since the first description of the 2DGE methodology, the technique has been used as a means to discriminate related isolates of the same bacterial species, for example *Neisseria* sp. [9, 10] and *Haemophilus* sp. [11, 12]. Detailed taxonomic studies have been described for *Listeria* sp. utilising 2DGE to analyse the bacterial cell proteins as a means to differentiate the isolates [13]. These early studies generally made no attempt to identify the resolved proteins and simple qualitative or quantitative differences in the protein patterns were scored to differentiate the bacteria. This comparative approach to proteomics has also been a key tool to look at bacterial pathogenesis and will be returned to later. The advent of methods to identify the individual proteins resolved by 2DGE permitted the development of detailed proteome databases linked directly to the genome sequence of the bacterium. Proteome databases based largely on proteins resolved by 2DGE are being developed for *Escherichia coli* [14, 15], *Bacillus subtilis* [16, 17] and the cyanobacterium, *Synechocystis* sp. strain PCC6803 [18, 19], each of which can be accessed via the Internet. Proteome databases are also being developed for clinically relevant bacteria and three examples (*Haemophilus influenzae*, *Helicobacter pylori* and *Mycobacterium tuberculosis*) will be considered below.

H. influenzae was the first free-living organism for which a complete genome sequence was obtained [20]. The *H. influenzae* proteome has been investigated by at least three groups [11, 12, 21–23]. However, The group of Fountoulakis and Langen has produced the most detailed picture of the *Haemophilus* proteome. The series of papers published by this group present an ideal strategy to dissect out the components of a bacterial proteome. The proteomics analyses have centred on the use of the Rd strain of *H. influenzae*, which is the avirulent laboratory strain used for the determination of the complete genome sequence. Direct analysis of cellular proteins by 2DGE, with no sample fractionation, resolved proteins in the pH range of 4–9.5 and mass range of 10–150 kDa [22]. This first pass of the proteome assigned identifications to 119 protein spots through the use of peptide mass mapping, amino acid composition analysis and *N*-terminal sequencing. As is well documented throughout proteomics this approach leads to an under-representation of low-abundance and hydrophobic proteins. Subsequent studies employed enrichment protocols prior to 2DGE to look for these 'problem' proteins. Heparin chromatography [24] and chromatofocusing [25] were used to enrich for low copy number proteins. Fractionation of bacterial cell lysates on heparin-actigel led to the enrichment of 160 proteins of which 110 were not detectable in unfractionated cell lysates [24]. Similarly, fractionation of the cell lysates by chromatofocusing led to the enrichment of 125 proteins of which 75 were novel [25]. In the case of *H. influenzae* the chromatofocusing provided the direct purification to homogeneity, or near homogeneity, of three bacterial proteins, major ferric iron-binding protein (HI0097), a hypothetical protein (HI0052) and 5′-nucleotidase (HI0206). There was no clear separation of the proteins solely on the basis of their pI by chromatofocusing with some pools collected from the columns containing proteins with pIs ranging between 5 and 9. The authors considered that co-elution of these proteins may have been due to the presence of protein complexes [25]. Neither of these chromatographic approaches enriched solely for the low copy number proteins and no single class of protein were selected by either method. Fountoulakis and Takacs [26] proposed the detailed cataloguing of the proteins enriched by specific chromatographic steps, for example heparin-actigel, which might be useful for the detailed analysis of specific protein classes. A similar spectrum of proteins is enriched by the same protocols from a range of different bacteria [26, 27] suggesting that data obtained with one microorganism can be extrapolated to others. Other classes of *H. influenzae* proteins have also been specifically analysed with the inclusion of basic (pI 6–11) [28] and low molecular mass (5–20 kDa) [29] proteins into their database. Both of these approaches modified the 2DGE separation technology itself to improve the analytical window available for protein separation. The combined data from these analyses have provided a detailed proteome map for *H. in-*

fluenzae with the assignment of 502 of the 1742 predicted proteins to the map and currently represents one of the largest microbial databases now available. This study also highlights the fact that no single method can reveal the entire proteome simultaneously and that complementary procedures are required to reveal low abundance proteins as well as proteins with extremes of pI and molecular mass.

H. pylori is a major cause of gastrointestinal infections and is of significant clinical interest since it has been implicated as a pre-disposing cause of gastric cancer. The genome is predicted to encode 1590 proteins [30] which is slightly larger than the *H. influenzae* genome but still within the scope of the current proteome technologies of 2DGE and peptide mass fingerprinting [31]. Jungblut et al. [32] have undertaken a detailed analysis of the *H. pylori* proteome and characterised the two sequenced *H. pylori* strains as well as a type strain used for animal studies. Up to 1800 protein species were resolved by 2DGE when silver staining was used to locate the proteins following electrophoresis. Peptide mass fingerprinting was used to identify 152 of the detected proteins for the 26695 strain; these represented 126 unique bacterial genes. Among the proteins identified were previously documented virulence determinants and antigens. Preliminary data also revealed the pH dependant expression of five protein spots. Due to the site of infection in the stomach the bacteria must survive extremes of pH and the detailed characterisation of proteins under acid conditions is certain to be a key area of investigation. At present the coverage of the proteome is still far from complete and these initial data are certain to be expanded. As with previous comparative studies [33, 34] of *Helicobacter* by 2DGE, Jungblut et al. [32] also found extensive variation at the level of the proteome.

The study of the *H. pylori* proteome has expanded beyond the purely gel based approach for its characterisation. Rain et al. [35] have presented a protein-protein interaction map for *H. pylori*, which provides connections between 46.6% of the proteins encoded by the proteome. Inspection of the protein interactions revealed by this approach is bringing to light specific biological pathways and contributes towards the prediction of protein function.

The proteome of *Mycobacterium tuberculosis* has been extensively characterised to look for potential markers of virulence as well as for specific changes induced during the interaction between bacteria and eukaryotic cells. These aspects of the proteome will be considered later. The following discussion considers the attempts to display the complete bacterial proteome. The genome of *M. tuberculosis* is 4,441,529 base pairs with a predicted coding capacity of 3924 genes. Two detailed studies of the *M. tuberculosis* proteome are underway [36–38] both of which can be accessed via the Internet. Jungblut and his colleagues [36, 37] have compared two virulent and two vaccine strains of *M. tuberculosis*. Up to 1800 and 800 proteins were resolved from either cell lysates or culture media respectively. Two hundred and sixty three protein spots were identified by peptide mass fingerprinting. This initial study has been expanded to look specifically at low molecular weight acidic proteins as part of a systematic analysis of the *M. tuberculosis* proteome [39]. Seventy six proteins, which have molecular masses between 6 and 15 kDa and migrated in the pH range 4–6, were excised and processed for peptide mass fingerprinting. Seventy two of the proteins were identified and were found to comprise about 50 structural proteins. In the second *M. tuberculosis* proteome database currently available Rosenkrands et al. [38] have looked at the H37Rv strain of *M. tuberculosis* and reported on the identification of 288 proteins present either in culture filtrates or cell lysates.

Urquhart et al. [40, 41] compared the predicted proteome map with that obtained experimentally from a series of six overlapping zoom 2D gels, which yielded 493 unique proteins, equivalent to approximately 12% of the expected protein coding. The predicted proteome map displays a bimodal distribution when plotting the predicted pIs and molecular masses for each protein [39, 41]. A similar bimodal distribution has also been observed for plots of the theoretical proteomes for *E. coli* [42] and *H. influenzae* [21]. Comparisons were made between the predicted and experimental determined proteome maps for *M. tuberculosis*. When the predicted and observed maps were superimposed outlier protein spots were observed in the observed map that were not present in the predicted proteome. These included 13 proteins with pIs <3.3, i.e. the lowest pI predicted from the genome sequence, as well as a cluster of low molecular mass (≤10 kDa) with pIs between 5 and 8. None of these outlier proteins were identified in the reported study, although it was suggested that they might have been derived through fragmentation or post-translational modification of the products of predicted open reading frames.

Jungblut et al. [43] have highlighted the capacity of proteomics to complement genome sequencing and in silico determination of genome coding for *M. tuberculosis*. Six genes were identified by proteomics (2DGE and peptide mass fingerprinting) that were not predicted from the genome sequence of the H37Rv strain. The identified proteins had molecular masses <11.5 kDa and migrated in the pH range 4.5–5.9. Partial sequencing by MS/MS of one of the proteins showed that the predicted DNA sequence derived from the peptide was present in the genome but that it had not been assigned in the original determination of predicted open reading frames. Similar observations have been reported by Rosenkrands et al. [44] who identified a 9-kDa (pI 4.9) protein that was not predicted from the known genome sequence.

3 Identification of Pathogenicity Determinants

An important research goal for many medical microbiologists is to understand how pathogenic bacteria interact with their hosts to produce clinical disease. Modern molecular technologies now allow researchers to probe these events in fine detail and, as discussed below, proteomics plays an important role in these developments.

3.1 Comparison of Virulent and Avirulent Bacterial Isolates

The ability of pathogenic bacteria to cause disease in a susceptible host is determined by multiple factors acting individually or together at different stages of infection. For example, two key virulence determinants, vacA and cagA, are involved in distinct aspects of *H. pylori* pathogenesis [45]; the cagA gene is itself a marker for a pathogenicity island containing approximately 29 ORFs [30]. The investigation of such complex phenomena on a gene-by-gene basis where the role of each gene is investigated in isolation is unreasonable. The availability of extensive gene sequence data and in some cases complete genome sequences for many bacterial pathogens opens up strategies for analysing global gene expression that are more appropriate for investigating polygenic phenomena like pathogenesis. The capacity of proteomics to analyse global protein synthesis and, by extension, gene expression in a non-specific manner makes this approach a powerful tool for identifying and characterising the expression of bacterial pathogenic determinants. A simple approach to investigate pathogenesis is to compare the proteins synthesised by virulent and avirulent bacterial strains, grown under standard conditions, in order to identify proteins that correlate with virulence. However, this approach has major limitations. First, when comparing naturally occurring bacterial variants, protein differences occur that are unrelated to virulence. Second, bacteria grown in vitro on defined culture media do not express all of the proteins encoded by the genome at levels characteristic of in vivo growth in the organism's natural host. This is apparent when comparing the proteins synthesised by facultative intracellular bacterial pathogens grown in defined culture media with the same bacteria growing in association with eukaryotic cells.

Despite these restrictions, high-resolution 2DGE is a popular method for identifying virulence determinants at the level of protein synthesis. Early studies comparing virulent and avirulent *Mycoplasma pneumoniae* isolates identified three novel proteins expressed by the virulent isolates that were absent from avirulent strains [46]. Comparisons of a virulent parental strain of *M. pneumoniae* with two derived avirulent mutant strains revealed both quantitative and qualitative differences in the protein profiles when analysed by 2DGE [47]. No data were provided on the identities of these proteins. A similar approach was used to compare virulent and avirulent vaccine strains of *Brucella. abortus* together with LPS deficient strains derived from each of the parental isolates [48]. Up to 935 proteins were resolved by 2DGE for the four strains. This was fewer than the expected 2129 proteins predicted based on the size of the 3.13×10^6-bp *B. abortus* genome [49]. The virulent and vaccine strains showed 98.4–99.3% homology at the DNA level. The amount of DNA equivalent to this difference in homology has a potential coding capacity of up to 34 proteins. The comparison of the 2D protein profiles of the virulent and avirulent strains identified 86 qualitative and 6 quantitative protein differences. Although the accumulation of point mutations in coding and regulatory sequences could have occurred, this alone would be insufficient to account for the observed difference in the DNA homology. An alternative proposal was that a genome rearrangement or deletion occurred that indirectly altered the expression of additional genes required to maintain cellular homeostasis [48].

Proteomics has been used to identify the virulence determinants of *Mycobacterium tuberculosis* by comparing virulent and vaccine BCG strains [36, 37, 40, 41, 50, 51]. Relatively few differences are detected at the proteome level for these bacteria [36] consistent with the limited genetic variability found from the sequence comparison of 26 genes among 842 *M. tuberculosis* complex isolates [52]. Urquhart et al. [40] used a high-resolution multi-gel system to compare the virulent *M. tuberculosis* H37Rv and *M. bovis* BCG (Pasteur ATCC 35734) strains. Up to 772 protein spots were identified for *M. bovis* BCG and virulent *M. tuberculosis* over a pH range of 2.3 – 11 with apparent molecular weights between 10 and 216 kDa. Some differences were observed between the two strains under these analytical conditions but their significance remains to be established. A more detailed comparison of virulent *M. tuberculosis* (H37Rv and Erdman) and vaccine (BCG Chicago and BCG Copenhagen) strains has been reported by Jungblut et al. [36]. The virulent and avirulent strains were grown in defined culture media and the bacterial proteins were extracted from both the cell lysates and culture supernatants. The majority of the proteins associated with the bacterial cells, as well as those released into the culture media, were common for all four isolates. The most extensive variation was found between H37Rv and BCG Chicago with 31 variant spots; 21 were qualitative and 10 were significant quantitative differences. There were also 18 and 3 protein spot variants observed in comparisons of the two virulent and two vaccine strains respectively. Although no novel virulence determinants were found for *M. tuberculosis* using proteomics, some virulence determinants identified by other means were assigned to the proteome [36]. The authors commented on the fact that amino acid substitutions leading to electrophoretic mobility variants might be useful for vaccine development if the substitution(s) occurred in T cell epitopes. Betts et al. [53] compared the proteomes of two virulent strains of *M. tuberculosis*, the laboratory adapted H37Rv strain and CDC1551, a recent clinical isolate [54] that has been partially sequenced. The analysis demonstrated that the classic virulent H37Rv strain used as the basis for many proteomic analyses had retained the features of more recent virulent *M. tuberculosis* isolates. A total of 1750 intracellular proteins were resolved by 2DGE over a range of pH 3 – 10 for each of the two strains examined when grown in vitro. Comparative studies of the proteomes of the bacteria assayed at various times during their growth revealed just 13 consistent spot differences between the isolates. Seven and three spots were specific for the CDC 1551 and H37Rv strains respectively. A further two spots were increased in abundance for H37Rv compared to CDC 1551. A single protein, identified as ribosome recycling factor, showed a mobility difference between the two isolates. Peptide mass mapping and MALDI-TOF were used to identify nine of the proteins exhibiting differences and identities were obtained. Four of the protein differences corresponded to mobility variants of MoxR (Rv1479); two variants were specific for each of the two bacterial strains. One of the CDC1551 specific proteins identified was a probable alcohol dehydrogenase (Rv0927c). One H37Rv specific proteins was identified as HisA and the two H37Rv induced spots were electrophoretic variants of alkyl hydroperoxide reductase chain C. The difference in the mobility of the MoxR protein was consistent with a nucleotide change observed in the gene for this protein. Consistent proteomes were observed for both isolates over the 12-day growth curve despite the fact that the CDC 1551 strain entered stationary phase in advance of H37Rv.

A difficulty in interpreting the data obtained from the comparative studies described above is in the selection of the specific virulent and avirulent strains used as the basis for the comparison. In many cases, the virulent and avirulent strains are genetically distinct and differences in their proteomes may be present that are unrelated to virulence. Mahairas et al. [50] used subtractive genomic hybridisation to locate genetic differences between *M. bovis* BCG vaccine and virulent isolates of *M. bovis* and *M. tuberculosis*. Three regions were deleted in the BCG vaccine compared to the virulent strains. One 9.5-kb segment (designated RD1) was absent from six BCG substrains but present in virulent *M. tuberculosis* and *M. bovis* strains as well as in 62 clinical *M. tuberculosis* isolates analysed. Based on its sequence RD1 contains at least eight open reading frames. The virulent *M. bovis* and *M. tuberculosis* showed indistinguishable protein profiles when compared by 2DGE. In contrast, the BCG vaccine strains expressed at least ten additional proteins and induced expression levels for a number of other proteins. When the RD1 region was introduced into the BCG genome to generate BCG::RD1 the protein profile of BCG::RD1 was indistinguishable from virulent *M. bovis*. This suggested that parts of RD1 caused a specific suppression of protein synthesis in virulent Mycobacteria. Some low molecular weight proteins were identified for BCG::RD1 that were consistent in size with short ORFs encoded in the RD1 sequence. More data are required to link these observations to the protein differences reported by Jungblut et al. [36] and Urquhart et al. [40] but once available a more complete picture of the Mycobacterial virulence determinants is sure to emerge.

3.2 In Vivo Induced Protein Synthesis

There are many limitations to identifying the determinants of bacterial pathogenesis by simply comparing bacterial isolates grown under laboratory conditions on defined culture media. A large number of studies have demonstrated that facultative bacterial pathogens express novel genes when they infect eukaryotic cells. Extensive efforts are being made by many research groups to identify these specific gene sets since they may represent novel pathogenic determinants or serve as potential targets for new therapeutic drugs. Recombinant DNA technologies, which take advantage of the existing genome sequence data now available, play a key role in identifying those bacterial genes specifically expressed in vivo, i.e. when the bacteria are in association with the eukaryotic cell. Reporter genes (for example Green Fluorescent Protein (GFP), β-galactosidase and luciferase) have been linked to bacterial promoters to monitor their in vivo induction (reviewed in [55]). An important experimental strategy that has been developed for looking at in vivo expressed genes is that of in vivo expression technology (IVET). The use of IVET can rapidly identify those bacterial genes that are expressed specifically in vivo either in cell culture systems or the intact animal. Proteomics provides a valuable complement to these DNA based technologies in looking at in vivo gene expression. However, there are severe restrictions in the use of proteomics for this field of research. Using current 2D gel based technologies proteomics is largely limited to analysing in vitro cell systems as described below. Using in vitro grown cell lines actually limits the type of data that can be derived. Although producing valuable and important data, in vitro grown cells do not fully mimic the intact animal where the bacteria are interacting at a number of levels with functionally distinct cell types present in an intact organism or tissue. At present there is no means to investigate protein synthesis of bacterial pathogens when grown in vivo with the equivalent sensitivity and power of IVET. The low recovery of bacterial cells from the host and an absence of a protein amplification method, analogous to the polymerase chain reaction for nucleic acids, present a number of technical difficulties to the investigator. Growth of the bacteria in artificial culture media, even for a limited time, is required which would invalidate the identification of in vivo specific protein synthesis. This may very well change in the near future once protein array technologies become generally available.

Nevertheless, progress can be made with in vitro model systems and proteomics has been widely used to examine the interaction of bacteria with eukaryotic cells. The use of in vitro cell lines has the advantage that they can be infected under reproducible controlled conditions and radioactive amino acids can be used to increase the sensitivity of protein detection. Typically, a combination of antibiotics and radioactive amino acids are used to selectively radiolabel the proteins synthesised by intracellular bacteria [56–61]. Briefly, eukaryotic cells are infected with bacteria and incubated with gentamycin to kill extracellular bacteria; the intracellular bacteria retain their viability. After a pre-determined time interval, radioactive amino acids are added in the presence of cycloheximide, which inhibits cellular but not bacterial protein synthesis. The radiolabelled proteins synthesised by the intracellular bacteria are compared with radiolabelled proteins prepared from bacteria grown in defined culture media The data obtained by this approach should be considered in the light of differences in the growth phase of the two bacterial populations as well as possible non-specific effects of the antibiotics used to inhibit cellular protein synthesis. An alternative to the use of cycloheximide, described by Burns-Keliher et al. [62], uses a modified lysine precursor to radiolabel *Salmonella typhimurium* proteins during infection of a human intestinal epithelial cell line. In this particular experimental system the precursor is specifically incorporated into the bacterial proteins but not mammalian cell proteins. The use of a cycloheximide block to inhibit cellular protein synthesis has been shown to be unnecessary for the analysis of *M. bovis* protein synthesis during infection of macrophage cells [63]. It was demonstrated that after radiolabelling in the absence of cycloheximide the macrophage cells could be lysed with SDS and the bacteria collected by centrifugation and washed with Tween-80. Under these conditions, control studies showed that there was minimal carry over of cellular proteins with the bacterial pellet. It was suggested that the ability to omit the cycloheximide allowed the cross signalling between the cell and bacteria which might otherwise be blocked by cycloheximide and so influence the pattern of bacterial growth within the cell [63].

Co-cultivation of bacteria with eukaryotic cells alone can be sufficient to induce the synthesis of proteins not expressed by bacteria grown in defined media. During the co-cultivation of Campylobacter jejuni with INT 407 cells, an epithelial cell line, at least 14 proteins showed increased biosynthetic levels. These changes in protein synthesis were revealed using 2DGE combined with either metabolic radiolabelling or immunoblotting [64, 65]. The induction of a subset of the 14 proteins also occurred following exposure of the bacteria either to cell culture medium alone or to INT 407 cell conditioned media [64]. It was suggested that the de novo synthesis of these proteins was required for the subsequent internalisation of the bacteria into the epithelial cells.

Intracellular bacteria growing in the cell's phagosome are exposed to a variety of stress conditions, including extremes of acidity, oxygen and nutrients [66]. In contrast, those intracellular bacterial pathogens that migrate out of the intracellular vacuoles into the cytoplasm are exposed to a reduced level of stress. During the infection of macrophage cells, *Legionella pneumophila, Brucella abortus* and *S. typhimurium* remain associated with the phagosome and may interfere with its maturation. A consistent observation for these intracellular bacterial pathogens is that the synthesis of specific bacterial proteins is either induced or repressed during the intracellular growth phase compared to bacteria growing in artificial culture media. Moreover, a number of the bacterial proteins induced during intracellular growth also show altered biosynthesis under in vitro stress conditions. The synthesis of the bacterial heat shock proteins GroEL and DnaK are induced during *B. abortus* infection of bovine macrophages [56]; the induced synthesis of GroEL has also been demonstrated in *B. abortus* infected murine macrophages [67]. The same two heat shock proteins also form part of the spectrum of proteins induced during *S. typhimurium* infection of macrophages [68]. Two major bacterial proteins homologous to the stress proteins DnaK and CRPA are induced during *Yersinia enterocolitica* infection of J774 cells, a murine macrophage cell line; these proteins are also induced in the bacteria by heat shock and oxidative stress in vitro [58]. Thirty two out of 67 bacterial proteins induced during *L. pneumophila* infection of the U937 macrophage cell line are also induced by in vitro stress conditions. These include the heat shock proteins GroEL and GroES [60]. A protein, global stress protein (GspA), is expressed by *L. pneumophila* in response to all stress conditions examined to date as well as during intracellular replication. GspA is induced at higher levels in intracellular bacteria suggesting that the bacteria in this particular environment may be exposed to multiple simultaneous stress conditions [66, 69]. Although the spectrum of proteins exhibiting altered synthesis by intracellular bacteria and bacteria stressed in vitro have many similarities, the changes observed for the former are not simply a summation of the in vitro stress responses [59] suggesting that these are specific responses induced during the intracellular replication phase.

Contrasting data to those described above are found for *Listeria monocytogenes* infection of J774 cells. A range of in vitro stress conditions, including heat shock and oxidative stress that induced the synthesis of GroEL and DnaK, induced none of the 32 proteins observed for intracellular bacteria [57]. The absence of known stress induced proteins expressed by the intracellular bacteria was believed due to the rapid migration of *Listeria* from the phagosome to the cytoplasm during intracellular growth [57].

M. bovis BCG infection of the THP-1 macrophage cell line results in the induced synthesis of bacterial proteins not expressed under standard in vitro growth conditions [63]. These proteins were demonstrated using radiolabelling as well as by immunoblotting against human *M. tuberculosis* infected sera. The induced immunogenic proteins may serve as future immunoprotective antigens.

Under the conditions used at least 20 proteins were differentially expressed in BCG-infected macrophages which were either specific for the infected macrophage cells or exhibited significantly induced levels of synthesis. Six of the proteins that were induced at 24 h post-infection were identified using a combination of MALDI and nanoES MS. These proteins were the GroEL homologues, GroEL-1/GroEL-2, InhA, 16-kDa antigen (α-crystallin Hsp-X), EF-Tu and a 31-kDa hypothetical protein. As commented upon by the authors these data are significant since they represent one of the earliest successful identifications of proteins recovered from intracellular bacteria using MS techniques [63].

As might be expected, the source of the eukaryotic cell can also influence the outcome of the bacterial infection. This has been shown for *M. tuberculosis* and *M. smegmatis* in which the type and origin of the cell used can influence both the initial interaction between bacteria and eukaryotic cell as well as the growth kinetics of the bacteria [70, 71]. The response of the bacterial proteome to bacterial growth in host cells of differing origin has been followed in *Mycobacterium avium* infection of bone marrow macrophages and J774 cells [72]. *M. avium* infection of J774 cells results in the specific induced synthesis of bacterial proteins. The induced bacterial protein synthesis commenced by 6 h post-infection and continued until at least four days post infection. None of the induced proteins showed altered synthesis when bacteria were exposed to in vitro stress conditions. In contrast to J774 cells, *M. avium* infection of primary bone macrophages followed different kinetics of bacterial replication and protein synthesis. Intracellular bacteria radiolabelled at 5 and 12 days post-infection showed significant differences in their protein synthesis. These data were consistent with the bacteria initially entering a stasis phase early in infection before commencing a normal replication cycle between 5 and 12 days post-infection. At 12 days post-infection the bacteria synthesised a similar range of proteins as *M. avium* grown in J774 cells. These observed variations in the replication cycle and protein synthesis of the bacteria using macrophage cells of different sources demonstrate the care required in selecting the host cell used to investigate intracellular bacterial replication at the level of the proteome.

The previous discussion considered the response of bacterial gene expression to the intracellular environment of the host. Similar approaches can, in principal, be used to investigate the host cell response to the bacterial infection. Investigations using proteomics as the principal method to examine the host response to bacterial infections at the cellular level have been fairly limited. A brief communication by Kovarova et al. [61] reported on global changes in macrophage protein synthesis following *Francisella tularensis* infection of mice. Principal component analysis was used to identify significant changes in protein synthesis between infected and uninfected cells. It was possible to distinguish uninfected and infected macrophage cells on the basis of their overall protein patterns obtained using 2DGE. The infected macrophage cells were further subdivided into macrophages collected three to seven days post-infection and cells collected at ten days post-infection [61]. No data were provided on the identities of the host proteins altered following infection. The *Nramp1* gene determines cellular resistance to intracellular parasites in mice [73]. *Nramp1* has a number of pleiotropic effects and 2DGE combined with principal component analysis has been used to identify those genes regulated by *Nramp1* [74]. Comparisons of protein synthesis in macrophages from mice carrying either the resistant or sensitive allele of *Nramp1* showed at least four proteins whose synthesis Nramp1 influenced. Two of the proteins with induced synthesis in macrophages carrying the resistant allele were provisionally identified as Mn-superoxide dismutase and bcl-2 [74]. Recently, Kovarova et al. [75] reported that expression of the resistance allele of *Nramp1* leads to modifications in the expression of a number of macrophage signal transduction pathways, which may be involved in providing resistance to microbial infection.

The effect of the infecting bacteria on the host cell proteome has also been examined for *M. tuberculosis*. The interaction of *M. tuberculosis* with the functions of the host cell phagosome has been extensively investigated using a number of criteria including at the level of the cellular proteome. During the intracellular replication of *M. tuberculosis* in macrophages the *M. tuberculosis* phagosomal compartments fails to fuse with the lysosomes, the normal fate of intracellular phagosomes. Essentially, the bacteria arrest the development and processing of the phagosome Comparisons of the phagosomal compartments from Mycobacterial infected cells with the same structures from uninfected cells have shown the occurrence of a number of differences in the cellular proteins present. In a recent review of the *M. tuberculosis* phagosome Fratti et al. [76] proposed that one of the key features of these structures was the exclusion of EEA-1 (early-endosomal autoantigen). This protein plays a role in vesicle tethering and also in endosomal fusion. At present the question remains on determining the identity of the protein, or proteins, expressed by *M. tuberculosis*, which leads to this exclusion of EEA-1.

4 Analysis of the Host Response to Bacterial Infection

4.1 Non-Specific Response to Bacterial Infection

The host can respond to the bacterial infection at a number of levels. One common non-specific response of eukaryotic hosts to infection by a variety of microbes including bacteria, fungi and protozoa is the withdrawal of iron from the infecting microbe. Iron is essential for the invading bacteria to permit multiplication. The host has developed a number of strategies for the withholding of iron from the invading microbes as described by Weinberg [77]. The acquisition of iron is also essential for virulence and this association has been widely reviewed (for example [78–82]). The acquisition of iron by bacteria is accomplished by the expression of small iron chelating molecules known as siderophores. For example, *M. tuberculosis* encodes two types of low molecular mass high affinity iron siderophores called exochelins and mycobactins. Exochelins are released into the extracellular the growth medium in large amounts whereas mycobactins remain associated with the bacterial cell wall [83–85].

Wong et al. [86] have investigated the occurrence of iron regulated gene expression in *M. tuberculosis* further. Growth of *M. tuberculosis* in vitro in either low (1 μm) or high (70 μm) iron concentrations resulted in 15 proteins showing an induced level of synthesis and 12 proteins with depressed synthesis for bacteria grown under low iron conditions. Under the analytical conditions employed in this study >250 cellular proteins were resolved by 2DGE following precipitation using 55–95% ammonium sulfate. Proteins with altered biosynthesis were identified using peptide mass mapping. Two well-documented iron-regulated proteins identified doing this study were Fur and aconitase. Fur, which was inhibited for bacteria growing in low iron concentrations, has been shown to be a regulator gene in a range of bacteria [87–89]. Aconitase was induced by high concentrations of iron and has a dual role in bacterial metabolism, functioning in the Krebs cycle and as an iron-responsive element binding protein. In high iron concentrations aconitase can form tight interactions with IREs in mRNA molecules involved in iron-storage in mammalian cells, for example transferin and ferritin. Other regulators that were induced in the presence of high iron concentrations include EF-Tu. In addition to known regulators of gene expression a variety of antigens (LSR-2, Hsp16.3) and enzymes (PEPCK, oxidoreductase and PPIase) exhibited differential expression under the iron concentrations.

4.2
Analysis of the Immune Response to Bacterial Infection

Progress has been achieved in the use of proteomics to investigate the host's immune response to infection. Two-dimensional electrophoresis combined with immunoblotting is widely used as a tool to investigate the humoral and cellular immune response against microbial pathogens. Immunoblotting against proteins separated by 1DE has been used for many years as a rapid means to compare the antibody profiles of human sera during or following infection [90–92] as well as for typing bacterial isolates on the basis of the electrophoretic mobilities of antigenic proteins [93, 94]. The analysis of bacterial proteins by 2DGE improves the resolution of antigenic proteins and increases the number of unique protein species amenable to analysis compared to one-dimensional separation methods. Combining this approach with protein identification by peptide mass mapping of the resolved proteins it is possible to identify rapidly the protein(s) reacting with the sera, an approach that may not always be possible using 1DE. Thus, one can improve significantly the identification of the spectrum of antibody specificities induced during infection.

The identification of immunogenic proteins by a combination of 2DGE and immunoblotting with polyclonal sera has been used for a number of bacteria, for example *Borrelia burgdoferi* [95], *Streptococcus pyogenes* [96] and *Brucella ovis* [97]. The global analysis of antigenic proteins using proteomics is a potentially useful approach to identify novel antigenic determinants for inclusion in future vaccines as initiated for *Helicobacter pylori*. Preparations of *H. pylori* cellular proteins were analysed by 2DGE and antigenic proteins identified by immunoblotting against a pool of human sera. Over 30 immunogenic protein spots were identified when the bacterial proteins were separated using IEF (pH 4–8) and NEPHGE for the first dimension separation of 2DGE. Seventeen immunogenic spots were located in the pH 4–8 region and a further 16 immunogenic protein spots in the pH 8–13 region [98, 99]. The antigenic proteins were characterised further by N-terminal sequencing and peptide mass mapping. Twenty-nine of the antigenic protein spots were identified and 15 were assigned to known ORFs in the *H. pylori* genome sequence; the remaining protein spots had homologies with bacterial proteins not present in the current *H. pylori* genome sequence. One 30-kDa protein spot contained two protein species both derived by post-translational processing from the *H. pylori* ORF, HP0175 [100] which had not previously been shown to encode an immunogenic protein. Comparisons of the proteins expressed by different isolates of *H. pylori* demonstrated that the majority of the antigenic proteins were expressed by all of the isolates, although differences in the expression levels of the flagellin protein and catalase were observed between some isolates [99].

Following the identification of the principal immunogenic bacterial proteins the next stage is to define the range of antibody specificities present in human sera for specific disease syndromes. Sanchez-Campillo et al. [101] determined antibody specificities against *Chlamydia trachomatis* proteins for seropositive patients with genital inflammatory disease. Antibodies were detected against a number of previously described antigenic proteins, including outer membrane proteins (OMP) and the GroEL-like and DnaK-like heat shock proteins. In addition, at least four novel immunogenic protein spots were detected that reacted with a variable proportion of the sera. The frequencies that these antigens were recognised varied among the patients, although all of the sera reacted against OMP2. As commented upon by these authors, a limitation in the data was the possible occurrence of cross-reactions with other Chlamydia species as well as unrelated bacteria [101]. Similar variations in the range of antibody specificities have been observed among patients with different clinical presentations of lyme borrelliosis [95]. In addition to the previously characterised *B. burgdorferi* antigens, the surface proteins and flagellin [102–104], some of the sera reacted with a number of undocumented immunogenic protein spots [95].

The identification of mycobacterial immunogenic proteins that elicit a protective cellular and humoral immune response during infection has received a lot of attention. The primary objective of these studies is the identification of novel immunogenic targets for new and improved vaccines. Previous studies to identify immunogenic bacterial proteins used 1DE in immunoblot assays against patients' sera [105, 106]. However, these methods failed to provide sufficient resolution to identify reliably the individual immunogenic proteins and improved resolution of the bacterial proteins was achieved using 2DGE. Antibody specificities have been examined among patients infected with *Mycobacterium leprae* who present with either borderline tuberculoid or lepromatous leprosy. Both patient groups have serum antibodies reacting against a major 30-kDa antigen [107]. Additional immunogenic protein spots were identified and at least six antigenic proteins were preferentially recognised by sera from patients with lepromatous leprosy. Peripheral blood lymphocytes prepared from leprosy patients and stimulated in vitro with anti-CD3 monoclonal antibody expressed antibodies that recognised a 10-kDa *M. leprae* protein. No antibodies reacting against this protein were detected in the sera collected from the same patients. The lymphocytes collected from healthy controls and processed in parallel reacted, possibly non-specifically, against some *M. leprae* proteins but not the 10-kDa protein. Although low molecular weight immunogenic proteins have been previously identified for *M. leprae* their relationship to the 10-kDa protein is unknown [107].

The T cell response forms a major component of the host's immune response against *M. tuberculosis* infection. The identification of those bacterial proteins that elicit the T cell response will significantly aid future vaccine development. It has been shown that, during growth in defined culture media, *M. tuberculosis* releases proteins into the media capable of inducing a protective cellular immune response [108–110]. Sonnenberg and Belisle [51] detected 205 protein spots from *M. tuberculosis* culture media by 2DGE. Thirty four unique proteins were identified in these preparations based on their reactions with specific monoclonal antibodies and amino acid sequencing. One member of the protein cluster at 83–85 kDa, previously recognised as a dominant humoral antigen [111], was identified as the *M. tuberculosis* KatG; at least two other members of this cluster shared an epitope with KatG [51]. The proteins were collected at 14 days post-infection, which is considered to be during the late logarithmic phase of bacterial growth when some bacterial cytoplasmic proteins may also be present in the media [51, 112]. Weldingh et al. [113] prepared cell-released proteins after only seven days growth of *M. tuberculosis* and analysed these by 2DGE. Six of the immunogenic proteins that were recognised by T cells were assigned to the Sanger *M. tuberculosis* genome sequence using N-terminal amino acid sequencing. One of the immunogenic proteins (designated CFP21) mapped onto the RD2 genome segment that is absent from some BCG strains [50]. A 29-kDa immunogenic protein present in the media from short-term cultures, as well as in the membrane of *M. tuberculosis*, is present in several mycobacterial species [114]. The KatG protein identified by Sonnenberg and Belisle [51] was not detected in these short-term culture filtrates.

The ability of proteins from *M. tuberculosis* infected culture filtrates and bacterial cells to stimulate T cells collected from tuberculosis patients, healthy controls as well as tuberculin positive and negative individuals has been examined [115, 116]. The bacterial proteins were analysed by 2DGE and then transferred to the liquid phase by electroblotting [117]. The individual protein fractions (400 from cell lysates and 480 from culture filtrates) were then used to stimulate T cells in vitro. Although variation existed between patients, T cells from tuberculosis patients and tuberculin positive individuals reacted with cell-associated antigens migrating at >30 kDa (pH range 4.2–6.6) [115]. These reaction patterns differed from those observed for the tuberculin negative individuals and healthy controls. Similar analyses for antigenic proteins in the culture filtrates revealed major antigenic components of 30–100 kDa and pIs of 4–5. T cells prepared from tuberculin negative contacts showed either a weak or no reaction to the bacterial proteins [116]. Although present in the culture media, neither the 38-kDa nor the 65-kDa major T cell antigens of *M. tuberculosis* migrated in this region [116]. More detailed studies [118] have recently been reported, which identify the antigenic proteins themselves. The fractions, collected as described previously [115], were assayed for T cell activation based on interferon-γ expression. Thirty eight fractions induced a significant T-cell response. The proteins present in these fractions were identified using trypsin digestion followed by LC-MS or LC-MS/MS to analyse the peptides. Positive identifications were achieved for 16 proteins from infected culture filtrates and 18 proteins from cell lysates. Although the majority of these proteins correspond to previously described antigens, 17 proteins were novel T-cell antigens.

The linkage of these diverse data on Mycobacterial immunogenic proteins will require the production of a detailed catalogue of cellular and cell-released proteins ideally linked to the developing *M. tuberculosis* proteomic databases [36, 37].

5 Determination of Therapeutic Strategies

In parallel with genome-orientated technologies proteomics has played a role in the development of novel therapeutic strategies via the identification of novel vaccine and antibiotic targets [119, 120]. Potential vaccine candidate proteins can be identified through the identification of in vivo immunogenic proteins. As discussed above antigenic proteins can be readily identified by characterising the immune response following either natural or laboratory infection with the bacterium of interest.

To optimise the identification of vaccine candidates, specific protein classes can be selectively analysed. Nilsson et al. [120] used a liquid phase charge separation to resolve putative membrane proteins of *H. pylori*. It is reasonable to expect that the membrane proteins are major targets for the host immune response. Bacterial cells were solubilised with *n*-octylglucoside and the proteins analysed with no further enrichment or fractionation. The bacterial cell proteins were fractioned by IEF in liquid phase for the 1st dimension and by continuous elution of selected IEF fractions by SDS-PAGE. Fifteen of the 40 proteins identified were membrane or membrane-associated proteins and many of these had not been previously identified using standard protocols of 2DGE. Chakravarti et al. [121] discussed a bioinformatics approach to the identification of vaccine candidate proteins. Their strategy takes advantage of the extensive gene sequence data that are now available. These authors used the *H. influenzae* Rd genome sequence to search for potential vaccine candidate proteins in silico. The identification of the candidate proteins was based on the detection of specific characteristics of outer membrane proteins since these are the most likely to be immunogenic and so most useful in vaccines. A complementary proteomic approach to identify the proteins from soluble outer membrane fractions of *H. pylori* was also described. Although this is a potentially valuable approach the authors did not confirm that the proteins identified by these criteria were in fact immunogenic under natural conditions.

The spread of antibiotic and drug resistance among microbial pathogens is a major problem for the control of infection. An understanding of the mechanism(s) by which drug resistance develops will lead to improvements in extending the efficacy of current anti-microbials. Proteomics can contribute towards determining anti-microbial resistance mechanisms through the capacity to analyse global changes in microbial proteins. Qualitative and quantitative changes can be identified in a non-specific manner without making pre-conceived judgements on the potential importance of different components.

Rifampin resistance in *Neisseria meningitidis* arises primarily through the mutation of the *rpoB* gene, which encodes the β subunit of the RNA polymerase [122]. Variants with a high level of resistance to rifampin can be selected by in vitro passage of *N. meningitidis* in the presence of rifampin. These variants show no additional mutations within the *rpoB* gene beyond that already demonstrated but there is evidence for an altered membrane permeability compared to the parental strains [123]. Analyses of the cellular proteins of the highly resistant and parental strains by 2DGE showed a shift to a more acidic isoelectric point for an 18.9-kDa protein for the highly resistant strains. Peptide mass mapping failed to find a match for this protein spot when searched against existing *Neisseria* sp gene sequences (L. Lawrie and P. Cash, unpublished data) and the significance of this protein mutation for either rifampin resistance or the altered membrane permeability remains to be determined.

Resistance to beta-lactam antibiotics has been investigated for *Pseudomonas aeruginosa*. A diminished expression of a 47-kDa (pI 5.2) outer membrane protein has been found for imipenem resistant *P. aeruginosa* [124]. Michea-Hamzehpour et al. [125] reported similar data with the loss of an outer membrane protein (pI 5.2) in imipenem resistant *P. aeruginosa*. *N*-Terminal sequencing showed that the protein was homologous to the porin outer membrane protein D [125]. Changes in outer membrane proteins have also been shown among ceftazidime resistant *P. aeruginosa* isolates with the expression of a basic protein homologous to the *ampC* gene product [125].

Penicillin resistance is on the increase among clinical isolates of *Streptococcus pneumoniae* and resistance to erythromycin, used as an alternative antibiotic to penicillin, has also emerged as a potential problem [126]. Two erythromycin resistant phenotypes are recognised for *S. pneumoniae*, specifically the MLS and M phenotypes [126]. Erythromycin resistant *S. pneumoniae* possessing the MLS phenotype owe their resistance to the methylation of rRNA by the product of the *erm* gene located on the transposon Tn*1545* [127, 128]. *S. pneumoniae* isolates with the M phenotype are less well-characterised, although resistance appears to be linked to the expression of a gene called *mefE* which is believed to encode a membrane transporter protein that reduces the intracellular levels of erythromycin [129, 130]. Proteomics has been used to investigate further the method of erythromycin resistance in M phenotype *S. pneumoniae* isolates [131]. Cellular proteins were prepared from erythromycin resistant (M phenotype) and sensitive isolates of *S. pneumoniae* and analysed using 2DGE. All nine M phenotype erythromycin resistant isolates analysed showed a characteristic induced synthesis of a 38.5-kDa protein (Fig. 1) [131]. None of the erythromycin sensitive *S. pneumoniae* strains showed this induced protein synthesis. The 38.5-kDa protein was identified as glyceraldehyde phosphate dehydrogenase (GAPDH) using peptide mass mapping and its peptide homology to GAPDH *S. equisimilus* and *S. pyogenes*. Three electrophoretic variants of GAPDH that differed in their pIs were resolved for M phenotype resistant isolates with the induced protein having the most basic isoelectric point. Thus, the abnormal synthesis of GAPDH in the M phenotype isolates may be related to an altered pattern of post-translational modification. The disrupted GAPDH synthesis was not related to erythromycin resistance per se since the MLS phenotype were indistinguishable

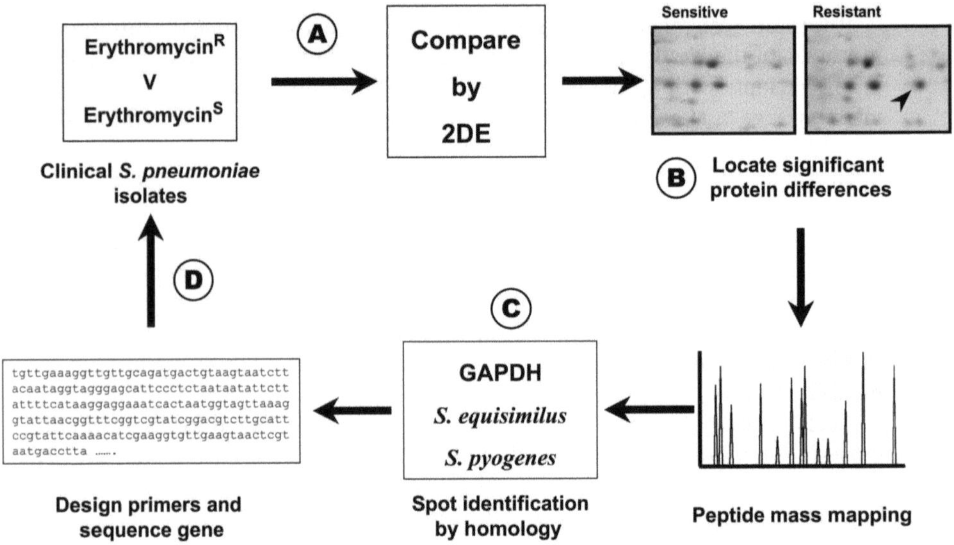

Fig. 1. Combined proteomic and genomic approach for characterising erythromycin resistance in *S. pneumoniae*. A collection of erythromycin sensitive and resistant *S. pneumoniae* isolates was compared by 2DGE (**A**) to locate protein differences. A characteristic difference was identified in the expression of a 38.5 kDa protein (**B**), which was identified as GAPDH by tryptic peptide mapping and its homology with GAPDH of *S. equisimilis* and *S. pyogenes* (**C**). A set of DNA primers were derived based on these gene sequences and used to sequence the gene encoding GAPDH in *S. pneumoniae* (**D**)

from sensitive *S. pneumoniae* isolates for GAPDH. Using the *S. equisimilus* and *S. pyogenes gap* gene sequences DNA primers were designed to analyse the GAPDH gene for the erythromycin sensitive and resistant *S. pneumoniae* isolates analysed by 2DGE [135]. A characteristic base mutation causing an amino acid substitution was shown to correlate completely with the protein profile demonstrated by 2DGE. This combination of proteomics and genomics to investigate erythromycin resistance is illustrated in Fig. 1 and clearly demonstrates the power of proteomics to locate putative gene targets for later study using genomic technologies.

The mechanism of antibiotic action can also be investigated by locating proteins that show differential expression patterns when bacteria are grown in the presence or absence of antibiotics. This experimental approach has been used to investigate isoniazid induced gene expression in *M. tuberculosis* using micro arrays [132]. The response of the bacterial proteome following their exposure to antibiotics has been investigated for *Staphylococcus aureus* [133]. *S. aureus* grown in the presence of inhibitory concentrations of oxacilin, a cell-wall active antibiotic, resulted in elevated expression for at least nine proteins as shown by 2DGE. Five of the induced proteins were identified by N-terminal sequencing as methionine sulfoxide reductase, enzyme IIA component of the phosphotransferase system, signal transduction protein (TRAP), GroES and GreA. A similar pattern of induced protein synthesis was found with other antibiotics that act on the bacterial cell wall but not by antibiotics acting on other targets, thus suggesting that the observed induced protein expression might represent a "proteomic signature" for this response [133]. A similar approach has been used to examine metronidazole induced gene expression in *H. pylori* [134]. Metronidazole resistance generally depends on the mutation of the *rdxa* (NADH reductase) gene with higher levels of resistance arising through mutations leading to the loss of function in additional reductase genes. When functional the reductase genes convert Metronidazole from a harmless drug to mutagenic and bactericidal products and, in the process, may generate reactive oxygen metabolites. Metronidazole resistant *H. pylori* were grown in sublethal concentrations of metronidazole and the differential protein expression determined by 2DGE. In presence of metronidazole 19 protein spots exhibited differential expression; 12 spots showed >2-fold decrease expression, 3 spots a >2-fold increased expression and no quantitative data were presented for the remaining 4 spots. Proteins that showed an increased expression level in the presence of metronidazole were identified by peptide mass mapping as alkylhydroperoxide reductase (AHP) (two protein spots) and aconitase B. AHP is know to protect against oxygen toxicity and it was proposed that the increased expression of AHP in metronidazole resistant *H. pylori* is important in the generation of the resistance phenotype.

6 Concluding Remarks

The experimental concepts that comprise the expanding field of proteomics are making rapid inroads into medical microbiology. Although it is unlikely to supplant fully recombinant DNA technologies in the study of pathogenic bacteria, proteomics provides a significant complementary approach. Many of the investigations presented in this review combine proteomics, genomics and bioinformatics to tackle the biological question.

As described in this review detailed proteome databases are being developed for clinically important bacteria, new data are being derived relating to pathogenic mechanisms and the host immune response. On the practical side there is the potential of developing novel targets for vaccines and therapeutic drugs. Thus progress is being made in many areas of study for pathogenic bacteria. There remains, however, the technical limitation of determining the proteomes of the bacteria when they are growing in association with their host. The in vitro cell studies described in this review simply scratch the surface of this aspect of bacterial pathogenesis. The analysis of bacterial growing in individual cell types do not fully replicate the diverse environmental conditions experienced by bacteria growing in the intact host with many distinct cell types not to mention a functional immune system. The technical limitation resides on the principal method that is used to resolve the proteins, namely 2DGE. The limitations of 2DGE with respect to sensitivity and problem protein classes have been well documented. The method is suitable for characterising bacteria grown under in vitro conditions as described above and 2DGE will remain a key technology in this area for a number of years. However, we need new methods with improved sensitivities and coverage to enable us to look at the bacterial proteome against the background of the host proteome. Time will tell if the new methods developed will be expanded from the current mass spectrometry methods or be the protein chips that are gradually being produced. Whichever method is ultimately developed it must have broad applicability, since not all research is carried out on fully sequenced model bacteria. Many of the new and exciting developments in medical microbiology will lie in studying recently isolated bacterial pathogens that are only poorly characterised at the molecular level.

7 References

1. Cash P (1998) Anal Chim Acta 372:121
2. VanBogelen RA, Greis KD, Blumenthal RM, Tani TH, Matthews RG (1999) Trends Microbiol 7:320
3. VanBogelen RA, Schiller EE, Thomas JD, Neidhardt FC (1999) Electrophoresis 20:2149
4. Gygi SP, Aebersold R (2000) Curr Opin Chem Biol 4:489
5. Holt LJ, Enever C, de Wildt RM, Tomlinson IM (2000) Curr Opin Biotechnol 11:445
6. Zhu H, Snyder M (2001) Curr Opin Chem Biol 5:40
7. Cordwell SJ, Wilkins MR, Cerpapoljak A, Gooley AA, Duncan M, Williams KL, Humphery-Smith I (1995) Electrophoresis 16:438
8. Cordwell SJ, Basseal DJ, Humphery-Smith I (1997) Electrophoresis 18:1335
9. Jackson P, Thornley MJ, Thompson RJ (1984) J Gen Microbiol 130:3189
10. Jackson P, Urwin VE, Torrance MR, Carmen JA (1989) Electrophoresis 10:456
11. Cash P, Argo E, Bruce KD (1995) Electrophoresis 16:135
12. Cash P, Argo E, Langford PR, Kroll JS (1997) Electrophoresis 18:1472
13. Gormon T, Phan-Thanh L (1995) Res Microbiol 146:143
14. VanBogelen RA, Abshire KZ, Pertsemlidis A, Clark RL, Neidhardt FC (1996) Gene-Protein database of *Escherichia coli* K-12, edn 6. In: Neidhardt FC, Curtiss R, Ingraham JL, Lin ECC, Low KB, Magasanik B et al. (eds) *Escherichia coli* and *Salmonella:* cellular and molecular biology. ASM Press, Washington, D.C., p 2067
15. Tonella L, Walsh BJ, Sanchez JC, Ou K, Wilkins MR, Tyler M, Frutiger S, Gooley AA, Pescaru I, Appel RD, Yan JX, Bairoch A, Hoogland C, Morch FS, Hughes GJ, Williams KL, Hochstrasser DF (1998) Electrophoresis 19:1960
16. Ohlmeier S, Scharf C, Hecker M (2000) Electrophoresis 21:3701
17. Hecker M, Engelmann S (2000) Int J Med Microbiol 290:123
18. Sazuka T, Ohara O (1997) Electrophoresis 18:1252
19. Sazuka T, Yamaguchi M, Ohara O (1999) Electrophoresis 20:2160
20. Fleischmann RD, Adams MD, White O, Clayton RA, Kirkness EF, Kerlavage AR, Bult CJ, Tomb JF, Dougherty BA, Merrick JM, McKenney K, Sutton G, FitzHugh W, Fields C, Gocayne JD, Scott J, Shirley R, Liu LI, Glodek A, Kelley JM, Weidman JF, Phillips CA, Spriggs T, Hedblom E, Cotton MD, Utterback TR, Hanna MC, Nguyen DT, Saudek DM, Brandon RC et al. (1995) Science 269:496
21. Link AJ, Hays LG, Carmack EB, Yates JR (1997) Electrophoresis 18:1314
22. Langen H, Gray C, Roder D, Juranville JF, Takacs B, Fountoulakis M (1997) Electrophoresis 18:1184
23. Langen H, Takacs B, Evers S, Berndt P, Lahm HW, Wipf B, Gray C, Fountoulakis M (2000) Electrophoresis 21:411
24. Fountoulakis M, Langen H, Evers S, Gray C, Takacs B (1997) Electrophoresis 18:1193
25. Fountoulakis M, Langen H, Gray C, Takacs B (1998) J Chromatogr A 806:279
26. Fountoulakis M, Takacs B (1998) Protein Expr Purif 14:113
27. Fountoulakis M, Takacs MF, Berndt P, Langen H, Takacs B (1999) Electrophoresis 20:2181
28. Fountoulakis M, Takacs B, Langen H (1998) Electrophoresis 19:761

29. Fountoulakis M, Juranville JF, Roder D, Evers S, Berndt P, Langen H (1998) Electrophoresis 19:1819
30. Tomb JF, White O, Kerlavage AR, Clayton RA, Sutton GG, Fleischmann RD, Ketchum KA, Klenk HP, Gill S, Dougherty BA, Nelson K, Quackenbush J, Zhou L, Kirkness EF, Peterson S, Loftus B, Richardson D, Dodson R, Khalak HG, Glodek A, McKenney K, Fitzegerald LM, Lee N, Adams MD, Hickey EK, Berg DE, Gocayne JD, Utterback TR, Peterson JD, Kelley JM et al. (1997) Nature 388:539
31. Bumann D, Meyer TF, Jungblut PR (2001) Proteomics 1:473
32. Jungblut PR, Bumann D, Haas G, Zimny-Arndt U, Holland P, Lamer S, Siejak F, Aebischer A, Meyer TF (2000) Mol Microbiol 36:710
33. Dunn BE, Perez-Perez GI, Blaser MJ (1989) Infect Immun 57:1825
34. Enroth H, Akerlund T, Sille A, Engstrand L (2000) Clin Diag Lab Immunol 7:301
35. Rain JC, Selig L, De Reuse H, Battaglia V, Reverdy C, Simon S, Lenzen G, Petel F, Wojcik J, Schachter V, Chemama Y, Labigne A, Legrain P (2001) Nature 409:211
36. Jungblut PR, Schaible UE, Mollenkopf HJ, Zimny-Arndt U, Raupach B, Mattow J, Halada P, Lamer S, Hagens K, Kaufmann SH (1999) Mol Microbiol 33:1103
37. Mollenkopf HJ, Jungblut PR, Raupach B, Mattow J, Lamer S, Zimny-Arndt U, Schaible UE, Kaufmann SH (1999) Electrophoresis 20:2172
38. Rosenkrands I, King A, Weldingh K, Moniatte M, Moertz E, Andersen P (2000) Electrophoresis 21:3740
39. Mattow J, Jungblut PR, Muller EC, Kaufmann SH (2001) Proteomics 1:494
40. Urquhart BL, Atsalos TE, Roach D, Basseal DJ, Bjellqvist B, Britton WL, Humphery-Smith I (1997) Electrophoresis 18:1384
41. Urquhart BL, Cordwell SJ, Humphery-Smith I (1998) Biochem Biophys Res Commun 253:70
42. VanBogelen RA, Abshire KZ, Moldover B, Olson ER, Neidhardt FC (1997) Electrophoresis 18:1243
43. Jungblut PR, Muller EC, Mattow J, Kaufmann SH (2001) Infect Immun 69:5905
44. Rosenkrands I, Weldingh K, Jacobsen S, Hansen CV, Florio W, Gianetri I, Andersen P (2000) Electrophoresis 21:935
45. Atherton JC (1998) Br Med Bull 54:105
46. Hansen EJ, Wilson RM, Baseman JB (1979) Infect Immun 24:468
47. Hansen EJ, Wilson RM, Clyde WA Jr, Baseman JB (1981) Infect Immun 32:127
48. Sowa BA, Kelly KA, Ficht TA, Adams LG (1992) Appl Theor Electrophor 3:33
49. Allardet-Servent A, Carles-Nurit MJ, Bourg G, Michaux S, Ramuz M (1991) J Bacteriol 173:2219
50. Mahairas GG, Sabo PJ, Hickey MJ, Singh DC, Stover CK (1996) J Bacteriol 178:1274
51. Sonnenberg MG, Belisle JT (1997) Infect Immun 65:4515
52. Sreevatsan S, Pan X, Stockbauer KE, Connell ND, Kreiswirth BN, Whittam TS, Musser JM (1997) Proc Natl Acad Sci USA 94:9869
53. Betts JC, Dodson P, Quan S, Lewis AP, Thomas PJ, Duncan K, McAdam RA (2000) Microbiology 146:3205
54. Valway SE, Sanchez MP, Shinnick TF, Orme I, Agerton T, Hoy D, Jones JS, Westmoreland H, Onorato IM (1998) N Engl J Med 338:633
55. Hautefort I, Hinton JC (2001) Philos Trans R Soc London - Ser B: Biol Sci 355:601
56. Rafie-Kolpin M, Essenberg RC, Wyckoff JH III (1996) Infect Immun 64:5274
57. Hanawa T, Yammamoto T, Kamiya S (1995) Infect Immun 63:4595
58. Yamamoto T, Hanawa T, Ogata S (1994) Microbiol Immunol 38:295
59. Abshire KZ, Neidhardt FC (1993) J Bacteriol 175:3734
60. Abu Kwaik Y, Eisenstein BI, Engleberg NC (1993) Infect Immun 61:1320
61. Kovarova H, Stulik J, Macela A, Lefkovits I, Skrabkova Z (1992) Electrophoresis 13:741
62. Burns-Keliher LL, Portteus A, Curtiss R III (1997) J Bacteriol 179:3604
63. Monahan IM, Betts J, Banerjee DK, Butcher PD (2001) Microbiology 147:459
64. Konkel ME, Cieplak W (1992) Infect Immun 60:4945
65. Konkel ME, Mead DJ, Cieplak W (1993) J Infect Dis 168:948

66. Kwaik YA, Harb OS (1999) Electrophoresis 20:2248
67. Lin J, Ficht TA (1995) Infect Immun 63:1409
68. Buchmeier NA, Heffron F (1990) Science 248:730
69. Abu Kwaik Y, Gao LY, Harb OS, Stone BJ (1997) Mol Microbiol 24:629
70. Mehta PK, King CH, White EH, Murtagh JJ, Quinn FD (1996) Infect Immun 64:2673
71. Barker K, Fan H, Carroll C, Kaplan G, Barker J, Hellmann W, Cohn ZA (1996) Infect Immun 64:428
72. Sturgill-Koszycki S, Haddix PL, Russell DG (1997) Electrophoresis 18:2558
73. Govoni G, Vidal S, Gauthier S, Skamene E, Malo D, Gros PP (1996) Infect Immun 64:2923
74. Kovarova H, Radzioch D, Hajduch M, Sirova M, Blaha V, Macela A, Stulik J, Hernychova L (1998) Electrophoresis 19:1325
75. Kovarova H, Necasova R, Porkertova S, Radzioch D, Marcela A (2001) Proteomics 1:587
76. Fratti RA, Vergne I, Chua J, Skidmore J, Deretic V (2000) Electrophoresis 21:3378
77. Weinberg ED (1995) Acquisition of iron and other nutrients in vivo. In: Roth JA, Bolin CA, Brogden KA, Minion FC, Wannemuehler MJ (eds) Virulence mechanisms of bacterial pathogens. ASM Press, Washington, D.C., p 79
78. Litwin CM, Calderwood SB (1994) J Bacteriol 176:240
79. Payne SM (1988) Crit Rev Microbiol 16:81
80. Finkelstein RA, Sciortino CV, McIntosh MA (1983) Rev Infect Dis 5:S759
81. Weinberg ED (1978) Microbiol Rev 42:45
82. Bullen JJ, Rogers HJ, Griffiths E (1978) Curr Top Microbiol Immunol 80:1
83. Barclay R, Ratledge C (1988) J Gen Microbiol 134:771
84. Sritharan M, Ratledge C (1989) FEMS Microbiol Lett 51:183
85. Gobin J, Moore CH, Reeve JR, Wong DK, Gibson BW, Horwitz MA (1995) Proc Natl Acad Sci USA 92:5189
86. Wong DK, Lee BY, Horwitz MA, Gibson BW (1999) Infect Immun 67:327
87. Lambert LA, Abshire K, Blankenhorn D, Slonczewski JL (1997) J Bacteriol 179:7595
88. Hassan HM, Sun HC (1992) Proc Natl Acad Sci USA 89:3217
89. Staggs TM, Perry RD (1992) Mol Microbiol 6:2507
90. Johnson PD, MacInnes SJ, Gilbert GL (1993) Infect Immun 61:1531
91. Agius G, Dindinaud G, Biggar RJ, Peyre R, Vaillant V, Ranger S, Poupet JY, Cisse MF, Castets M (1990) J Med Virol 30:117
92. Gimenez HB, Keir HM, Cash P (1987) J Gen Virol 68:1267
93. Morgan MG, McKenzie H, Enright MC, Bain M, Emmanuel FX (1992) Eur J Clin Microbiol Infect Dis 11:305
94. Wolfhagen MJ, Fluit AC, Torensma R, Jansze M, Kuypers AF, Verhage EA, Verhoef J (1993) J Clin Microbiol 31:2208
95. Jungblut PR, Zimny-Arndt U, Zeindl-Eberhart E, Stulik J, Koupilova K, Pleissner KP, Otto A, Muller EC, Sokolowska-Kohler W, Grabher G, Stoffler G (1999) Electrophoresis 20:2100
96. Lemos JA, Giambiagi-Demarval M, Castro AC (1998) J Med Microbiol 47:711
97. Teixeira-Gomes AP, Cloeckaert A, Bezard G, Bowden RA, Dubray G, Zygmunt MS (1997) Electrophoresis 18:1491
98. McAtee CP, Fry KE, Berg DE (1998) Helicobacter 3:163
99. McAtee CP, Lim MY, Fung K, Velligan M, Fry K, Chow T, Berg DE (1998) Clin Diagn Lab Immunol 5:537
100. McAtee CP, Lim MY, Fung K, Velligan M, Fry K, Chow TP, Berg DE (1998) J Chromatogr B Biomed Sci Appl 714:325
101. Sanchez-Campillo M, Bini L, Comanducci M, Raggiaschi R, Marzocchi B, Pallini V, Ratti G (1999) Electrophoresis 20:2269
102. Krause A, Burmester GR, Rensing A, Schoerner C, Schaible UE, Simon MM, Herzer P, Kramer MD, Wallich R (1992) J Clin Invest 90:1077
103. Mathiesen MJ, Hansen K, Axelsen N, Halkier-Sorensen L, Theisen M (1996) Med Microbiol Immunol 185:121
104. Batsford S, Rust C, Neubert U (1998) J Infect Dis 178:1676

105. Chakrabarty AK, Maire MA, Lambert PH (1982) Clin Exp Immunol 49:523
106. Ehrenberg JP, Gebre N (1987) Scand J Immunol 26:673
107. Mahon AC, Gebre N, Nurlign A (1990) Int Immunol 2:803
108. Pal PG, Horwitz MA (1992) Infect Immun 60:4781
109. Roberts AD, Sonnenberg MG, Ordway DJ, Furney SK, Brennan, PJ, Belisle JT, Orme IM (1995) Immunology 85:502
110. Andersen P (1994) Infect Immun 62:2536
111. Laal S, Samanich KM, Sonnenberg MG, Zolla-Pazner S, Phadtare JM, Belisle JT (1997) Clin Diagn Lab Immunol 4:49
112. Andersen P, Askgaard D, Ljungqvist L, Bennedsen J, Heron I (1991) Infect Immun 59:1905
113. Weldingh K, Rosenkrands I, Jacobsen S, Rasmussen PB, Elhay MJ, Andersen P (1998) Infect Immun 66:3492
114. Rosenkrands I, Rasmussen PB, Carnio M, Jacobsen S, Theisen M, Andersen P (1998) Infect Immun 66:2728
115. Schoel B, Gulle H, Kaufmann SH (1992) Infect Immun 60:1717
116. Daugelat S, Gulle H, Schoel B, Kaufmann SH (1992) J Infect Dis 166:186
117. Gulle H, Schoel B, Kaufmann SH (1990) J Immunol Meth 133:253
118. Covert BA, Spencer JS, Orme IM, Belisle JT (2001) Proteomics 1:574
119. Rosamond J, Allsop A (2000) Science 287:1973
120. Nilsson CL, Larsson T, Gustafsson E, Karlsson KA, Davidsson P (2000) Anal Chem 72(9):2148
121. Chakravarti DN, Fiske MJ, Fletcher LD, Zagursky RJ (2000) Vaccine 19:601
122. Carter PE, Abadi FJ, Yakubu DE, Pennington TH (1994) Antimicrob Agents Chemother 38:1256
123. Abadi FJ, Carter PE, Cash P, Pennington TH (1996) Antimicrob Agents Chemother 40:646
124. Vurma-Rapp U, Kayser FH, Hadorn K, Wiederkehr F (1990) Eur J Clin Microbiol Infect Dis 9:580
125. Michea-Hamzehpour M, Sanchez JC, Epp SF, Paquet N, Hughes GJ, Hochstrasser D, Pechere JC (1993) Enzyme Protein 47:1
126. Johnson AP, Speller DC, George RC, Warner M, Domingue G, Efstratiou A (1996) BMJ 312:1454
127. Clewell DB, Flannagan SE, Jaworski DD (1995) Trends Microbiol 3:229
128. Trieu-Cuot P, Poyart-Salmeron C, Carlier C, Courvalin P (1990) Nucleic Acids Res 18:3660
129. Tait-Kamradt A, Clancy J, Cronan M, Dib-Hajj F, Wondrack L, Yuan W, Sutcliffe J (1997) Antimicrob Agents Chemother 41:2251
130. Sutcliffe J, Tait-Kamradt A, Wondrack L (1996) Antimicrob Agents Chemother 40:1817
131. Cash P, Argo E, Ford L, Lawrie L, McKenzie H (1999) Electrophoresis 20:2259
132. Wilson M, DeRisi J, Kristensen H, Imboden P, Rane S, Brown PO, Schoolnik GK (1999) Proc Natl Acad Sci USA 96:12,833
133. Singh VK, Jayaswal RK, Wilkinson BJ (2001) FEMS Microbiol Lett 199:79
134. McAtee CP, Hoffman PS, Berg DE (2001) Proteomics 1:516
135. Amezaga MR, Carter PE, Cash P, McKenzie H (2000) The 2nd International Symposium on Pneumococcal Diseases, S. Africa

Received: April 2002

Adv Biochem Engin/Biotechnol (2003) 83: 117–140
DOI 10.1007/b11114

Application of Proteomics to *Pseudomonas aeruginosa*

Amanda S. Nouwens · Bradley J. Walsh · Stuart J. Cordwell

Australian Proteome Analysis Facility, Sydney, Australia 2109
E-mail: s.cordwell@proteome.org.au

The recent completion of the *Pseudomonas* Genome Project, in conjunction with the *Pseudomonas* Community Annotation Project (PseudoCAP) has fast-tracked our ability to apply the tools encompassed under the term 'proteomics' to this pathogen. Such global approaches will allow the research community to answer long-standing questions regarding the ability of *Pseudomonas aeruginosa* to survive diverse habitats, its high intrinsic resistance to antibiotics and its pathogenic nature towards humans. Proteomics provides an array of tools capable of confirming the expression of Open Reading Frames (ORF), the relative levels of their expression, the environmental conditions required for this expression and the sub-cellular location of the encoded gene-products. Since proteins are important cellular effectors, the biological questions we pose can be defined in terms of changes in protein expression detectable by separation to purity using two-dimensional gel electrophoresis (2-DGE) and relation to gene sequences via mass spectrometry. As such, we can compare strains with well-characterized phenotypic differences, growth under a variety of stresses, protein interactions and complexes and aid in defining proteins of unknown function. While the complete genome has only recently been finished, a number of studies have already utilized this information and examined various protein gene-products using proteomics. This review summarizes the application of proteomics to *P. aeruginosa* and highlights potential areas of future research, including overcoming the traditional technical limitations associated with 2-DGE. More focused approaches that target sub-cellular fractions ('sub-proteomes') prior to 2-DGE can provide further functional information. A review of current and previous proteomic projects on *P. aeruginosa* is presented, as well as theoretical considerations of the importance of sub-proteomic approaches to enhance these investigations.

Keywords. Membrane proteins, Culture supernatant, Mass spectrometry, Two-dimensional gel electrophoresis, Sub-proteome, Strain comparison

List of Abbreviations

2-DGE	Two-dimensional gel electrophoresis
CF	Cystic fibrosis
IPG	Immobilized pH gradient
MALDI-TOF MS	Matrix-assisted laser desorption ionisation time-of-flight mass spectrometry
OMP	Outer membrane protein
ORF	Open reading frame
PMM	Peptide mass mapping
PTM	Post-translational modification
QS	Quorum sensing
TMR	Transmembrane spanning region

1 Introduction

Pseudomonas aeruginosa is a bacterium regarded as ubiquitous in the environment, living in soil, water and vegetation. Its unusual ability to proliferate in such diverse and extreme environments is surpassed only by its pathogenicity towards humans. As an opportunistic pathogen, the organism infects immuno-compromised individuals such as those with cystic fibrosis, burns victims and cancer patients (reviewed in [1]). *P. aeruginosa* is the second leading cause of nosocomial infections in the United States. These infections include pneumonia, endocarditis and infections of the eyes, ears, skin, urinary tract and central nervous system. At present, the clinical outcome for approximately 50% of hospitalized cystic fibrosis, cancer and burn patients with a pseudomonad infection is mortality. This situation is compounded by the increasing resistance *P. aeruginosa* displays to many classes of antibiotics, including beta-lactams and aminoglycosides [2]. Furthermore, the large genetic coding capacity (approximately 5500 genes) of the organism has allowed it a degree of genome plasticity that enables it to evolve rapidly mechanisms to resist typical commercial detergents and soaps, and thus persist in consistently cleaned environments, such as hospitals.

1.1 Pathogenicity of *Pseudomonas aeruginosa*

The ability of *P. aeruginosa* to initiate and establish infection in a host is facilitated by the production of assorted factors, including proteins that are located on the cell surface or secreted into the external environment. Well-characterized examples of these factors are listed in Table 1. Pathogenic factors act in a variety of ways during the infectious process to protect the bacterium and allow it to proliferate in the host. While for many organisms proteins are the sole molecules involved in this process, *P. aeruginosa* contains a variety of other, non-proteinaceous, substances that are involved in pathogenicity. Extracellular factors include pigments such as pyocyanin, which can induce reactive oxygen species in host cells [3], pyoverdine which scavenges iron and is implicated in cystic fibrosis [4, 5] and other compounds such as hydrogen cyanide that inhibits cytochrome C oxidase [6]. *P. aeruginosa* also secretes a large array of proteins that are significant in pathogenicity. These include proteases such as elastase (LasB) and alkaline protease that degrade proteins on the surface of host cells and tissues [7],

Table 1. Known pathogenicity factors produced by *P. aeruginosa*

Gene	Protein	Sub-cellular location	Virulence function
lasA	LasA protease	Extracellular	Elastolytic and proteolytic activity [7, 62, 76–78]
lasB	Elastase	Extracellular	Elastolytic and proteolytic activity (reviewed in [7])
exoA	Exotoxin A	Extracellular	ADP-ribosylation preventing protein synthesis (reviewed in [1])
exoS	Exoenzyme S	Extracellular	ADP-ribosylation preventing protein synthesis [79, 80]
phnA, phnB	Pyocyanin	Extracellular	Increases reactive oxygen species intracellularly [81]
Many genes	Lipo-polysaccharide	Cell-surface	Protection from phagocytosis [82–84]
algR and	Alginate	Cell-surface	Epithelial cell ligand; protection from host others defenses [85, 86]
fliC	Flagella	Cell-surface	Establish infection, interact with epithelial membranes [1, 87, 88]
pchC, pchE, pchF	Pyochelin	Extracellular	Siderophore: iron-chelating. Increases reactive oxygen species [89]
pvcA-E	Pyoverdine	Extracellular	Siderophore: iron-chelating [4, 17]
aprA	Alkaline protease	Extracellular	Protease; anticoagulant activity
plcH, plcN	Phospholipase C	Extracellular	Hydrolysis of phosphatidylcholine; possible compromise neutrophil function [90]
pilD, pilS, pilR	Type 4 pili	Cell-surface	Surface adhesion [14]

and toxins such as exotoxin A which are necrotising to host tissue [8]. Many of the virulence factors produced by *P. aeruginosa* are simply secreted outside of the bacterium into the external environment, while others, such as exoenzyme S (ExoS), are specifically directed or "injected" into host cells [9] by the type III secretion system. Extracellular virulence factors are primarily expressed at or near stationary phase, with many identified virulence factors under the control of one or both of the quorum-sensing systems so far identified in *P. aeruginosa* (discussed in detail later).

While secreted proteins play an important role in regulating virulence, *P. aeruginosa* maintains an arsenal of cell surface components that are involved in adherence, cell-cell signalling, transport of nutrients (typical Gram negative porin-like proteins) and pathogenicity, as well as resistance to a variety of chemicals and antibiotics. The pseudomonal outer membrane is highly impermeable to foreign compounds and those that can penetrate intracellularly may be removed via a series of protein efflux pump complexes (reviewed in [10]) designated as Mex-Opr complexes. These are three component systems encoding an inner membrane antiporter, an efflux OMP (Opr), and a fusion protein that maintains the stability and activity of the complex. A regulatory repressor (encoded by *mexR*) has recently been found [11, 12], and mutations in this gene are associated with increased antibiotic resistance [13]. *P. aeruginosa* maintains surface mechanical structures including flagella, pili and fimbriae, and these have been implicated not only in motility, but also in adherence and thus virulence. For example, type IV pili act as a mediator for interactions between the bacterium and the host [14], as well as playing a role in cytotoxicity [15]. Finally, *P. aeruginosa* responds to nutrient limitation by releasing a set of scavenging siderophores. For example, under iron-limitation the siderophore pyoverdin is released into the extracellular milieu and has a high affinity for free iron, which it scavenges from the host. The pyoverdin-iron complex is then bound by a specific outer membrane receptor (e.g. FpvA; [16, 17]) that facilitates the transport of scavenged iron into the cell.

2
The *Pseudomonas* Genome and Community Annotation Project

2.1
Genome Sequencing and Annotation

The genome era has revolutionized the way in which biologists design their experiments. Rather than taking a single gene approach, the tools now exist to perform genome- or proteome-wide analyses. The plethora of genetic information derived from genome sequencing projects can provide, for example, clues as to the way an organism survives and proliferates under various environmental conditions or a subset of potential vaccine or diagnostic candidates. The complete genome sequence for any organism should be treated as a tool for interrogation leading to sensible experimental design. The complete genome of *P. aeruginosa* was recently sequenced (www.*pseudomonas*.com) and this has provided much new information that will better our understanding of this bacterium and facilitate discovery of new targets for antimicrobial agents and/or vaccines [18]. Access to the *P. aeruginosa* genome sequence and annotations can also be made via The University of Queensland database (pseudomonas.bit.uq.edu.au; [19]), or for specific analysis of outer membrane proteins, cmdr.ubc.ca/bobh/genomics.htm.

The Pseudomonas Genome Project determined the genome to be 6.37 Mb in size and encompassing over 5500 open reading frames (ORFs), making it the largest bacterial genome sequenced to date. This remarkable coding capacity may be reflected in the large number of predicted proteins belonging to the two-component signal transduction family. Such complex genetic regulation and the large coding capacity of the organism may provide it with a mechanism to adapt rapidly to new environments and thus develop resistances to many chemicals and antibiotics. To facilitate the correlation of gene sequence to predicted function, an annotation consortium, the *Pseudomonas* Community Annotation Project (PseudoCAP), was established and is comprised of expert volunteers in the field [20]. Such a system was unique in bacterial genome annotation and allowed scientists with a high degree of functional knowledge of the biochemistry and physiology of the organism to test their predictions against the genetic code. Annotations were defined through comparison with known genes that had been previously functionally characterized in *P. aeruginosa*, or via sequence similarity with genes from other bacterial species. Nearly 50% of the ORFs were found to have no known function, being either unique to *P. aeruginosa* or having sequence similarity to 'hypothetical' proteins from other bacteria. Clearly, functional annotations based on sequence similarity with ORFs in other organisms are not as dependable as those with demonstrated biochemical function in *P. aeruginosa* itself. Hence a limited range of confidence levels continue to be assigned by PseudoCAP to each ORF annotation to help differentiate those known, well-characterized genes and gene-products from those with little known function. Interestingly, of the 5570 ORFs, only 372 (6.7%) currently have the highest confidence level possible, i.e. their function has been experimentally determined in *P. aeruginosa.* Obviously, much molecular biology and microbial biochemistry remain to elucidate the in vivo functions of the majority of the genome.

As many ORFs do not yet have a demonstrated role in *P. aeruginosa*, there is much interest in determining these functions. *P. aeruginosa* does not maintain a significantly higher proportion of paralogous genes, suggesting that gene duplication has not contributed alone to the complexity of the genome. Instead, it appears the additional ORFs may represent genes involved in biochemical pathways unique to *P. aeruginosa*, and that allow the organism to perform secondary metabolic functions. It has been suggested that these pathways possibly allow it to scavenge nutrients from specific or unusual sources (e.g. xenobiotic compounds), which in turn allows *P. aeruginosa* to proliferate in many diverse environments. Furthermore, *P. aeruginosa* is thought to be among the most 'evolvable' prokaryotes yet discovered [21], giving it the capacity to adapt readily to novel surroundings. Clearly, with so many ORFs being unique to *P. aeruginosa*, global studies of gene and protein expression under an almost infinite number of conditions will need to be undertaken.

2.2 The Need for Proteomics and Transcriptomics in Genome Projects

While the genome sequence has recently been completed, and has provided information such as genome size, %G+C composition, total number of ORFs, homologous comparisons of ORFs to those in other organisms, etc., it cannot provide functional information that is dependant on monitoring gene expression under a given set of conditions, or on post-translational modifications of those expressed protein gene-products. Genome sequencing cannot indicate which genes are specifically induced or the levels of induction or expression under a chosen condition. Analysis of genomic information alone provides only a static view of the genes present and the potential pathways needed for survival and proliferation, but not demonstrative evidence for their role in vivo. Tools based on a variety of computational methods and algorithms, such as similarity searches, certainly have a role in predicting gene function, but are not always reliable, and may result in errors in functional assignment of genes, especially where the sequence similarity is weak. Protein folding and tertiary structure also play a significant role in determining function.

Functional analysis of gene expression can be conducted via many approaches, including examination of the mRNA and protein gene-products, alone or in combination with experimental approaches such as gene knockout mutagenesis, nutrient limitation, adaptation to stress and host-pathogen interactions. Microarray technology (transcriptomics; reviewed in [22]) allows the level of mRNA transcripts to be examined at a global level in response to chosen biological conditions. Transcriptional mRNA studies have been applied to examine the effect of *pilA* mutants on human lung epithelia during infection [23], but have not yet been used to observe directly *P. aeruginosa* gene expression. DNA microarrays specific for *P. aeruginosa* are now becoming commercially available, while others have described methodology for generating in-house arrays (Proceedings of Pseudomonas 2001, Brussels, Belgium). Whole genome microarrays have been used to detect changes in gene expression associated with quorum-sensing in *Escherichia coli* [24] and biofilm formation in *Vibrio cholerae* [25]. Both experiments have implications for processes known to be performed by *P. aeruginosa* and thus similar experiments in this organism will provide a framework for comparative studies across species boundaries. Although mRNA studies provide one approach to understanding gene function, mRNA is only another step in the gene-to-product path and transcript levels do not necessarily reflect protein (translation) levels in the cell [26]. Despite the suggestion that there is better correlation between mRNA and protein levels in bacterial species compared to eukaryotes, co- and post-translational modifications, as well as stability and half-life are not necessarily reflected at the mRNA level. Without doubt a complete analysis of global gene expression under a given biological condition is only possible through the combination of transcriptomics and proteomics approaches.

3 Proteomics

The term 'proteomics' describes the global separation, characterization and analysis of proteins expressed by a genome, as well as the "tools" involved (reviewed in [27, 28]). In recent times the field has expanded to include differential protein display (traditionally two-dimensional gel electrophoresis and mass spectrometry), protein-protein interactions via the yeast 2-hybrid system or affinity chromatography and protein structure-function studies. Proteomics incorporates methodologies capable of confirming gene sequences (particularly start and stop sites), the relative levels of expression, environmental conditions required for gene expression and the sub-cellular location of these gene-products. Currently, the most common approach to a global examination of proteins is separation of complex mixtures via two-dimensional gel electrophoresis (2-DGE). At this time, no other method is available for separating so many individual components to a relative purity that enables post-separation analysis. Alternate methods based on liquid chromatography [29], or with quantitation provided by isotope-coded affinity tags (ICAT; [30, 31]) are now being developed to allow separation and identification of proteins without 2-DGE. However, at this stage neither method has been applied to *P. aeruginosa* and thus will not be discussed here.

Advances in 2-DGE, such as stable, reproducible immobilized pH gradients (IPGs), the incorporation of improved solubilising and reducing agents (e.g. tributyl phosphine [32]), synthetic zwitterionic detergents including amido-sulfobetaine 14 (ASB-14; [33]) and improved detection methods, such as fluorescent dyes (e.g. Sypro Ruby) [34], now allow reproducible separation of proteins within a variety of physical parameters. While 2-DGE technologies have been improved, the 'proteome era' has been made possible only with the rapid advances in peptide mass spectrometry that now allows rapid correlation of a gel-purified protein with a database gene sequence (reviewed in [35, 36]). These technological advances, combined with genomic information, have resulted in a new surge of interest in 2-DGE and associated proteomic tools for the analysis of *P. aeruginosa* proteins.

3.1 Whole Proteomes and Sub-Proteomes

In spite of the above-mentioned recent improvements, limitations still exist in proteomic analyses, such as the limited loading capacity of IPGs and the total area of separation available on 2-D gels. Furthermore, there are groups of proteins, including hydrophobic, basic and low abundance proteins, that are not amenable to 2-DGE using standard protocols. The 'dynamic range' of protein expression, in particular, makes the resolution of all proteins in a single experiment particularly difficult. To overcome this, pre-fractionation has been employed in an attempt to provide a 'third dimension' to 2-DGE analyses. Such improvements have included sequential extractions [37], where proteins are differentially solubilised across a series of stronger detergents and chaotropes, and separation across varying narrow or 'zoom' pH ranges in the first dimension [38]. These methods essentially fractionate proteins based on an inherent physical property, such as relative solubility and/or p*I*, providing a set of less complex sub-proteomes (defined as a fraction of the entire proteome). Furthermore, sub-proteomics can also be extended to the examination of a particular cell fraction, organelle or sub-cellular location. Using *P. aeruginosa* PAO1 as an example, we have previously shown how a combination of differential solubilising steps and narrow-range pH gradients can be used to resolve proteins better, with up to a 45% increase in the observed proteome [38]. Most proteome-style projects conducted on *P. aeruginosa* thus far have concentrated on proteins from whole cells, resolved using a one-step solubilising buffer, and a broad range pH gradient for separation. While these projects are biologically significant, and have provided functional information about proteins for which there is limited information, there is a distinct possibility that other functionally significant proteins are excluded due to the limited resolving capacities of the chosen 2-DGE parameters, particularly alkaline or hydrophobic proteins and low copy number proteins swamped by abundant proteins (dynamic range). Admittedly, application of multiple solubilising steps and multiple gels with overlapping gradients is technically challenging and may not always be necessary, particularly for comparative analyses. Our own experience with multiple solubilising steps on *P. aeruginosa* is that, although some novel proteins can be extracted in subsequent fractions compared to the initial extraction step (due to increased solubilising power of the 2-DGE sample buffer), the majority of proteins will appear in more than one fraction.

A better alternative for sub-proteomic approaches is fractionation based on functional properties related to the organism, such as the relative cellular location (e.g. membrane, extracellular, cytosolic, and for eukaryotic projects, additional sub-cellular fractions such as mitochondrial, nuclear, ribosomal fractions, etc.). Fractionation based on sub-cellular location is an approach particularly useful for aiding in the characterization of hypothetical or unknown proteins as greater functional information can be obtained. Sub-cellular fractionation also results in reduced 2-DGE map complexity as extraneous or irrelevant proteins to the particular study can be excluded. This in turn also allows a greater quantity of the proteins of interest to be applied to the 2-D gel. We have used this sub-cellular approach to examine membrane proteins from *P. aeruginosa* [39] in which the membranes are first isolated and precipitated using a high pH sodium carbonate buffer, and the membrane proteins extracted in a strong solubilising buffer containing the new synthetic detergent ASB-14 (Fig. 1) [33]. Isolation and solubilisation of proteins from *P. aeruginosa* outer membranes allowed the resolution of approximately 300 protein spots of which 189 were subsequently identified using mass spectrometry. From this, 16% had no known function, and another 46% had defined function based solely on sequence similarity to genes from other organisms. The isolation of those proteins annotated as "hypothetical" or "unknown" in the membrane fraction suggests that they are, in fact, membrane-associated. We found no contamination from known cellular proteins in the membrane preparation aside from the ubiquitous 60-kDa chaperonin GroEL, which is one of the most abundant proteins expressed in *P. aeruginosa*, and potentially shuttles to and from the membrane in its role as a chaperone. Therefore,

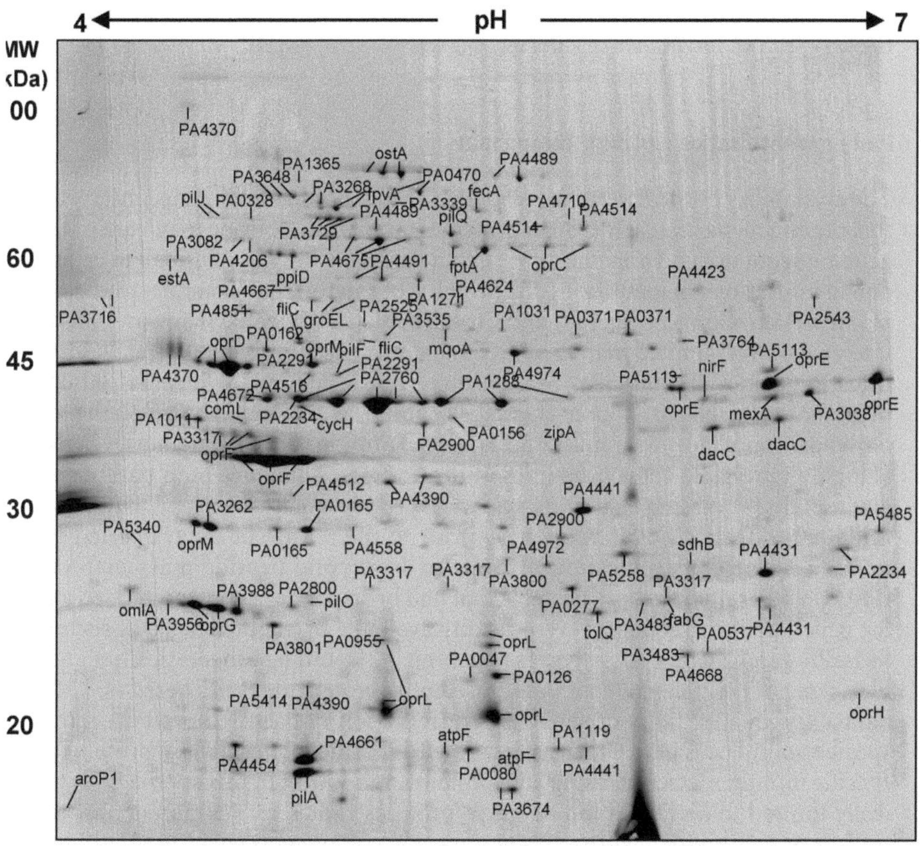

Fig. 1. Reference map of outer membrane proteins from *P. aeruginosa*. Membrane proteins were enriched with a sodium carbonate treatment (Fig. 3; [53]) and separated on pH 4–7 2-D gels. Protein spots were characterized by PMM, and protein identities assigned by comparison with the translated PA01 genome sequence. Protein spots are labelled with gene names or *P. aeruginosa* (PA) accession numbers

it is most likely that the 28 proteins of unknown function determined in this study are novel OMPs that have not previously been characterized in any microorganism. This subset is thus highly amenable to further study, especially in the search for new drug targets. The methodologies defined in this study have allowed the approach to be transferred to the examination of OMPs across strains and during host-pathogen interactions in vivo.

Other researchers have also utilized a sub-proteomic approach to examine the outer membrane and cell surface proteins from *P. aeruginosa*, in particular to determine differential gene expression in response to biological conditions. Multiple differences were detected in *P. aeruginosa* membrane protein composition either in response to the presence of antibiotics [40] or due to nutrient limitation [41]. With the powerful improvements in 2-DGE and mass spectrometry we have described already, it is anticipated that several such studies will begin in the coming years and provide many avenues for targeted research aimed at elucidating the function of molecules of interest.

3.2 Protein Identification and High-Throughput

The availability of the *P. aeruginosa* genome sequence has heightened interest in this organism, particularly from a proteomic perspective. Prior to release of the genome sequence and annotation, proteomic approaches were difficult to utilize, due to limitations in identifying the proteins of interest. For example, studies by Michea-Hamzehpour et al. [40] and Cowell et al. [41] detected many proteins of interest but were limited in their ability to identify these proteins accurately. Identification relied on sufficient protein quantity to attempt *N*-terminal Edman degradation sequencing. Some characterization was conducted based on the relative positions (molecular mass and p*I*) of the proteins on 2-D gels; however, this is highly unreliable since protein modifications can influence both parameters and protein processing resulting in cleavage products may also occur. In our own initial investigations of *P. aeruginosa* prior to the release of the translated genomic sequence, identification of proteins by tryptic peptide mass mapping (PMM) was limited to those proteins that had been previously characterized and for which protein sequences were readily available in public databases such as SwissProt and NCBI. Cross-species matching [42], although theoretically a good idea, was not practical for confidently assigning protein identities based on PMM data alone due to the poor sequence conservation of bacterial proteins across species boundaries [43]. Other projects have identified *P. aeruginosa* proteins by specific immunodetection using anti-protein antibodies, but this approach is typically limited to identification of single proteins rather than a more global approach, or using patient antisera where specific components eliciting an immune response cannot be identified within the sample complexity. However, such immune responsive proteins are often secreted or are present on the cell surface.

The availability of the total genetic information now makes it possible to identify many more proteins expressed by *P. aeruginosa*, given the appropriate amounts of material and access to the corresponding techniques for protein characterization. With access to the genomic information, peptide mass mapping via matrix-assisted laser desorption ionisation/time-of-flight mass spectrometry (MALDI-TOF MS) has become the fastest method-of-choice for protein identification. In conjunction with automated procedures including spot cutting and chemical dispensing for gel plug washing, trypsinization and elution, MALDI target-spotting, data acquisition and database searching, 200 protein spots or more can be characterized by a single MALDI-TOF MS in a single working day. Software programs such as MassLynx (Micromass, Manchester UK) aid in the automation of PMM spectrum acquisition, data calibration and database searching with minimal user intervention. Figure 2 shows the data output for one example from *P. aeruginosa* using MassLynx, including the mass spectrum, potential matches and sequence coverage for the match. However, great care must be taken to realize that while such a capacity is useful, it does not, nor should not, replace sensible experimental design aimed at finding a very small subset of

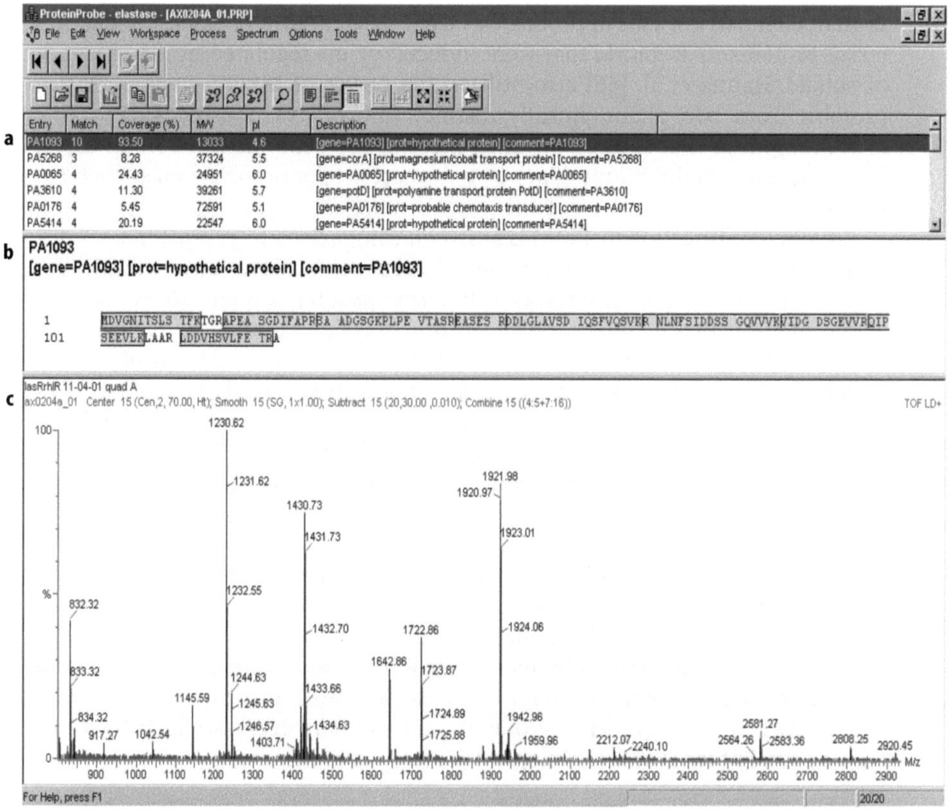

Fig. 2a–c. Screenshots of database output from PMM protein identification using automated ProteinLynx software. PMM data from MALDI-TOF MS was compared to the translated PA01 genome database: **a** potential database matches are sorted by protein sequence coverage (%); *highlighted* gene PA1093, encoding a hypothetical protein, with 93.5% protein sequence coverage and 10 matching peptides corresponds to the identified protein; **b** a visual display of protein sequence coverage provides additional confidence to the assigned match; *highlighted* sequence indicates those amino acids matched by PMM; **c** the mass spectrum is also displayed to allow comparison of peak intensity with peaks matched in the PA01 database

proteins associated with a biologically relevant process. Where PMM is unsuccessful a hierarchical approach is taken to protein identification that includes peptide concentration and desalting, differential enzymatic cleavage and electrospray ionisation (ESI) tandem MS [44]. This approach is more reliable than traditional Edman sequencing approaches since more than one peptide sequence can be obtained. *N*-terminal analyses maintain a significant role in determining protein start sites, especially where biological cleavage events are thought to occur.

Although *N*-terminal sequencing can conclusively identify proteins, modern approaches such as MS/MS de novo sequencing also allow protein identification by peptide sequencing and require only a fraction of the material required for

Edman chemistries. This approach was used by Quadroni et al. [45] to characterize proteins in *P. putida* that were induced or up-regulated in the presence of sulfate. Hanna et al. [46] also utilized the power of MS/MS to characterise proteins from two phenotypically distinct (mucoid vs non-mucoid) strains of *P. aeruginosa* that showed no differences at the genomic level. In all cases, however, research directions depend on the access to appropriate mass spectrometers.

Protein identification by PMM is based on comparison of peptide masses observed in the mass spectrometer to theoretical peptide masses of enzymatically-digested protein sequence databases. *P. aeruginosa* has a high %G+C content which suggests that it codes for a lower total of lysine than other organisms. In comparison, *Mycobacterium tuberculosis* is the only sequenced microbe to have a higher %G+C content and it is estimated that approximately 17% of its ORFs encode proteins with no tryptic peptides in the 800–3500 Da mass 'window' usually detected in PMM [47]. Therefore, it is possible that many *P. aeruginosa* proteins will be difficult to analyse using standard tryptic PMM. Furthermore, it is also apparent that database sequences are not always entirely correct, even before consideration of events such as the removal of signal sequences or post-translational modification (including protein cleavage). Given these considerations, however, for an organism such as *P. aeruginosa* a good signal to noise ratio in the PMM mass spectrum should result in a confident match to a single ORF in the genome. In our experience, the main difficulties in matching quality spectra to an ORF in the genome has been for low molecular mass proteins and those that represent cleavage/degradative products, such as found in extracellular fractions where protein turnover and biological proteolysis are common.

3.3
Application of Proteomics to *P. aeruginosa*

Proteomics can be applied in one of two main ways: to identify all proteins expressed under a chosen set of conditions ('reference mapping'), and/or as a means of comparing differential protein expression from one or more environmental or genetic variations. Reference maps obviously form a basis from which all proteins under the chosen conditions can be identified and then used as landmarks for comparative work. Ultimately the usefulness of reference mapping is relative to the number of protein spots that can be confidently identified (see [44] for review). Reference maps of whole cell extracts can provide information regarding the number of proteins, potentially the extent of protein modifications (through gene:gene-product ratios) and the relative abundance of the expressed proteins. However, unless combined with a pre-fractionation step, such as subcellular fractionation, reference maps can be of limited value, as little functional information, other than expression, regarding unknown and conserved hypothetical proteins is obtained. Comparative approaches, however, allow identification of differentially expressed proteins (e.g. induced or up-regulated or alternatively down-regulated or switched off) in response to growth or genetic variations.

3.4 Membrane Proteins

In bacteria, membrane and cell surface proteins play a vital role in cell survival, proliferation and virulence in the host, and as such have been widely investigated in *P. aeruginosa* via a number of approaches, including proteomics. Membrane and cell surface proteins function to promote adherence to surfaces (e.g. pili), and in *P. aeruginosa* are particularly important as efflux pumps to remove antibiotics (e.g. MexAB-OprM), as well as receptors and channels for essential nutrients (e.g. vitamin B12 receptor and sugar-transporting porins) [48, 49]. Aside from interest in membrane proteins to understand further the physiological aspects of *P. aeruginosa*, membrane or surface proteins are also of interest as potential epitopes for vaccine development. Much work has been performed showing the efficacy of purified outer membranes, specific OMPs, or representative OMP-derived peptides in the quest for a *P. aeruginosa* vaccine [50–52]. However, like all organisms, membrane proteins from *P. aeruginosa* have been somewhat difficult to resolve by 2-DGE. Improvements in solubilising agents, such as the synthetic detergent ASB-14 [33] have been shown to improve both qualitatively and quantitatively, the resolution of membrane-associated proteins for a number of Gram-negative bacteria, including *P. aeruginosa* (Fig. 1; [39]) and *E. coli* [53], as well as eukaryotic organisms [54]. A schematic of the methodology used to separate cytosolic, outer membrane and extracellular sub-proteomes is shown in Fig. 3. However, it should also be noted that many outer membrane proteins are hydrophilic throughout their sequence, yet contain one or more hydrophobic transmembrane spanning regions (TMR). We have visualized proteins with up to 7 TMR [39], while others have viewed eukaryotic proteins with 12 TMR [54]. Of the 189 proteins identified in our analysis of the pseudomonal outer membrane, only 11 had theoretical Kyte-Doolittle/Grand Average Hydropathy (GRAVY) values above zero. This indicates that proteins with even low overall hydrophobicity remain difficult to solubilise for proteomic analyses. Despite traditional beliefs, SDS does not appear to solubilise a significantly greater number of these proteins when used in conjunction with sequential extractions (unpublished data). Some of the earliest proteomic work on *P. aeruginosa* was based on membrane proteins [40, 41]. Solubilisation of membrane proteins in these instances utilised either CHAPS or SDS, and as such, the separation and resolution of these molecules was limited, either due to the poor solubility of the membrane or due to ionic disturbances caused by charged detergents interfering in the isoelectric focusing. In our analyses of *P. aeruginosa* OMPs, we identified 104 unique ORFs amongst 189 characterised protein spots. The functions of these proteins could be divided into three wide categories – (I) porins, (II) receptors and (III) unknown function. The biggest group consisted of 17 proteins with significant sequence similarity to porins (Porins C, D, E1, F, H1, OprM, H8 and OprF). Image analysis showed that five of these ORFs (OprD, E, F, G and potential OmpE3) accounted for the expression of over 50% of the total protein visible on 2-D gels of *P. aeruginosa* outer membranes. These porins are generally specific for the transport of nutrients; for example OprD is specific for basic amino acids including glutamine and arginine, small basic peptides

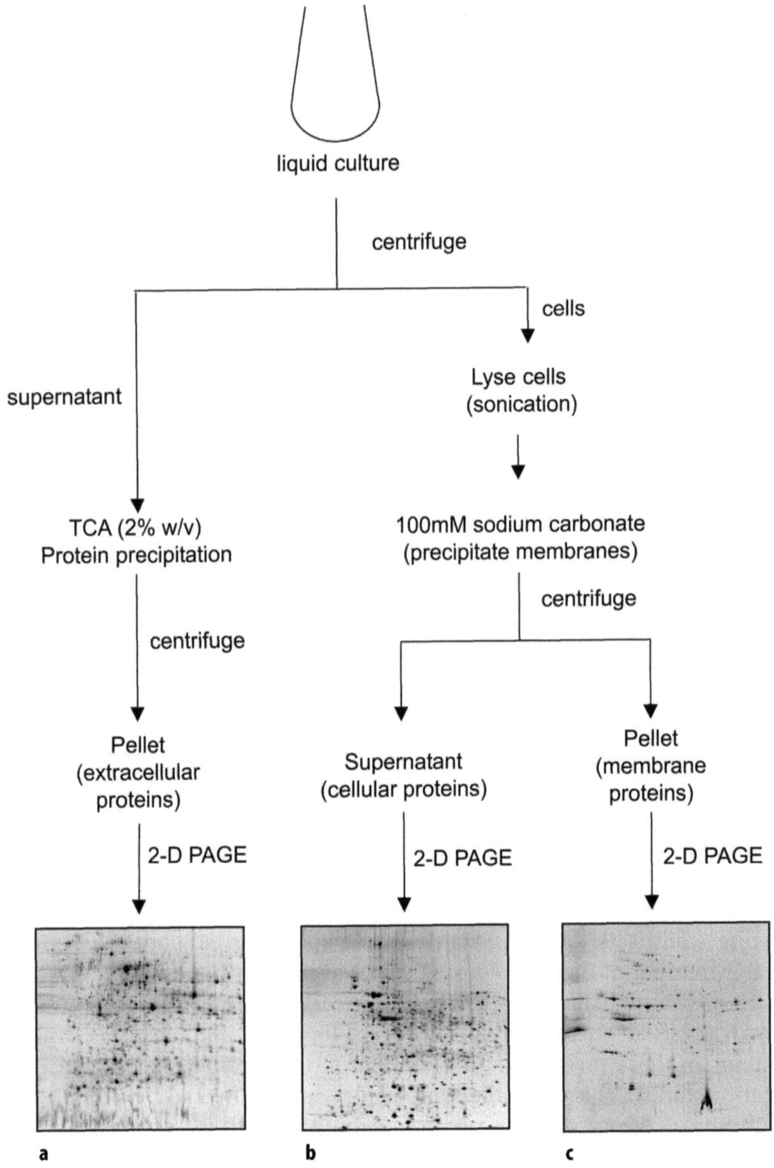

Fig. 3. Proteomic approach for bacterial protein analysis. Schematic showing the isolation of proteins from the extracellular, membrane and cytoplasmic fractions. Liquid cultures are centrifuged and extracellular proteins **a** isolated from the supernatant by precipitation with trichloroacetic acid/methanol. Cellular proteins **b** can be isolated by incubation in sodium carbonate followed by centrifugation. The pellet contains the membrane proteins **c**, while the cytoplasmic proteins in the supernatant can be precipitated from the carbonate buffer with trichloroacetic acid/methanol. Suitable buffers for protein solubilisation for extracellular and cytoplasmic proteins: 5 mol/l urea, 2 mol/l thiourea, 2% CHAPS, 2% sulfobetaine 3–10, 1% carrier ampholytes, 2 mmol/l tributyl phosphine; and for membrane proteins: 7 mol/l urea, 2 mol/l thiourea, 1% amidosulfobetaine-14, 1% carrier ampholytes, 2 mmol/l tributyl phosphine

and by analogy the beta-lactam antibiotic imipenem [55]. The presence of amino acids including arginine and histidine in the medium as sole carbon or nitrogen sources induce the expression of OprD suggesting a complex metabolic role for this protein [56]. OprF is the major pseudomonal outer membrane protein and although it is defined as a non-specific porin, it appears to have multiple functions including the maintenance of cell shape and association with peptidoglycan [57, 58]. This protein has also been the basis for both DNA and protein vaccines [51, 59]. In the second category, receptors for several siderophores, including pyochelin, pyoverdine and aerobactin were identified. The expression of such proteins may be constitutive, or indeed the cells may have been iron-starved even in the complex media used to perform the study. Since OMPs play such a vital role in pathogenicity, the methodology now developed to routinely examine these proteins via global proteomics analysis should provide a means to disclose rapidly new functions when combined with biological experimentation.

3.5 Extracellular Proteins

As *P. aeruginosa* is a pathogen for humans, there is also much interest in secreted proteins, many of which function as virulence factors (Table 1). Protein secretion in *P. aeruginosa* is dependent, in part, on growth phase with the majority of extracellular proteins expressed at or near stationary phase. Additionally, many of the genes encoding extracellular proteins are regulated by quorum sensing [60].

Extracellular proteins are relatively easy to isolate as they are generally hydrophilic and thus easily solubilised. However, at present little proteomic work has been conducted on culture supernatant (CSN) proteins from *P. aeruginosa*. Characterisation and identification of these proteins has previously been based on SDS-PAGE and Western blotting to probe for antigens that elicit an immune response. We have been conducting proteomic analyses of culture supernatant proteins (Fig. 3) from phenotypically different strains of *P. aeruginosa*, and also examining the influence of quorum sensing on extracellular protein expression. Reference maps of extracellular proteins derived from the PA01 strain during exponential and stationary phase growth are shown in Fig. 4. *P. aeruginosa* secretes proteins into the extracellular environment in response to signals that include quorum-sensing of population density and nutrient limitation. The major secreted protein of *P. aeruginosa* is elastase (LasB), which has a broad range of specificity allowing it to target host tissues and immune systems. LasB is highly immunogenic and antibodies against this protein have also been found in cystic fibrosis patients [61]. Another endopeptidase, LasA protease, is also a major component of the *P. aeruginosa* culture supernatant. This protease is specifically directed towards multiple glycine or gly-gly-ala sequences in proteins, providing it with staphylolytic activity due to the glycine cross-linkages in cell wall peptidoglycan. Several other minor constituents were observed in stationary phase culture supernatants; however, one further major protein was identified as a 'hypothetical aminopeptidase' that has now been fully character-

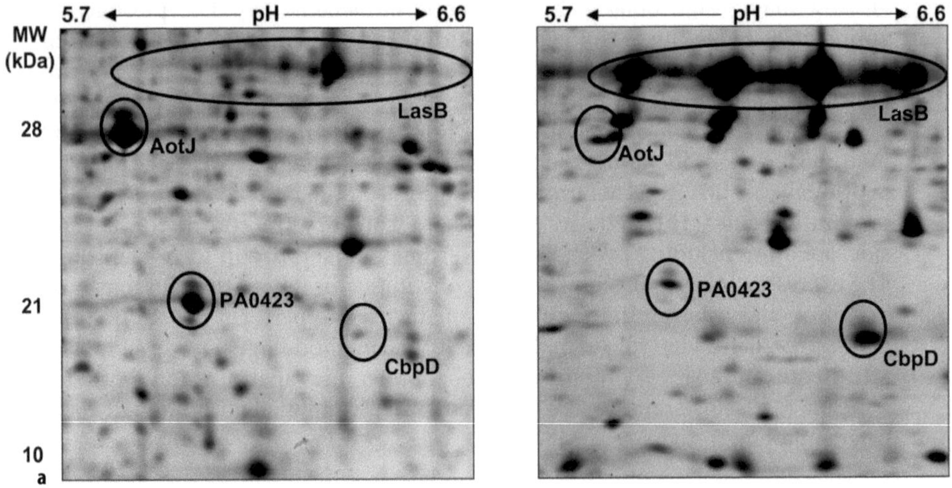

Fig. 4a, b. Comparison of extracellular proteins from *P. aeruginosa*: **a** exponential; **b** stationary phase cultures. Proteins were visualized with SyproRuby and characterized by PMM. Protein spots are labelled with gene names. Significantly greater amounts of LasB (elastase), and CbpD (chitin-binding protein) are expressed during stationary phase, while less AotJ (arginine-ornithine binding protein) and PA0423 are present

ized [62]. Other constituents of the *P. aeruginosa* culture supernatant include sheared flagellar proteins such as FliC and FliD and breakdown products of surface-accessible OMPs.

We have also undertaken a global analysis of quorum sensing (QS) mutants using proteomics of *P. aeruginosa* extracellular fractions (Nouwens et al., in preparation). QS is the mechanism used by bacteria to communicate population density and is a complex regulatory system that allows the organism to respond as a group by controlling gene expression. The QS system in *P. aeruginosa* (encoded by *lasI/lasR* and *rhlI/rhlR;* Fig. 5) is thought to regulate the expression of almost 5% of the genome [63] including a number of virulence factors [64, 65]. As shown in Fig. 5, many of these are controlled solely by the *las* or *rhl* operons; however the expression of others, including LasB itself, can be regulated by either system in the absence of the other. Virulence factors regulated by QS include elastase (*lasB*), LasA protease (*lasA*), alkaline phosphatase (*phoA*), and hemolysin, as well as non-protein extracellular molecules, such as pyocyanin. In our experiments, multiple deletion mutants consisting of single, double and quadruple gene disruptions were analysed via 2-DGE and MS. Several novel QS-regulated genes were identified (unpublished data). Figure 6 shows the effects of these mutations on the levels of expression on one well-characterized QS-regulated protein (LasA protease). Proteomics, coupled with transcriptomics, will be able to shed much light on the mechanisms of QS in *P. aeruginosa.*

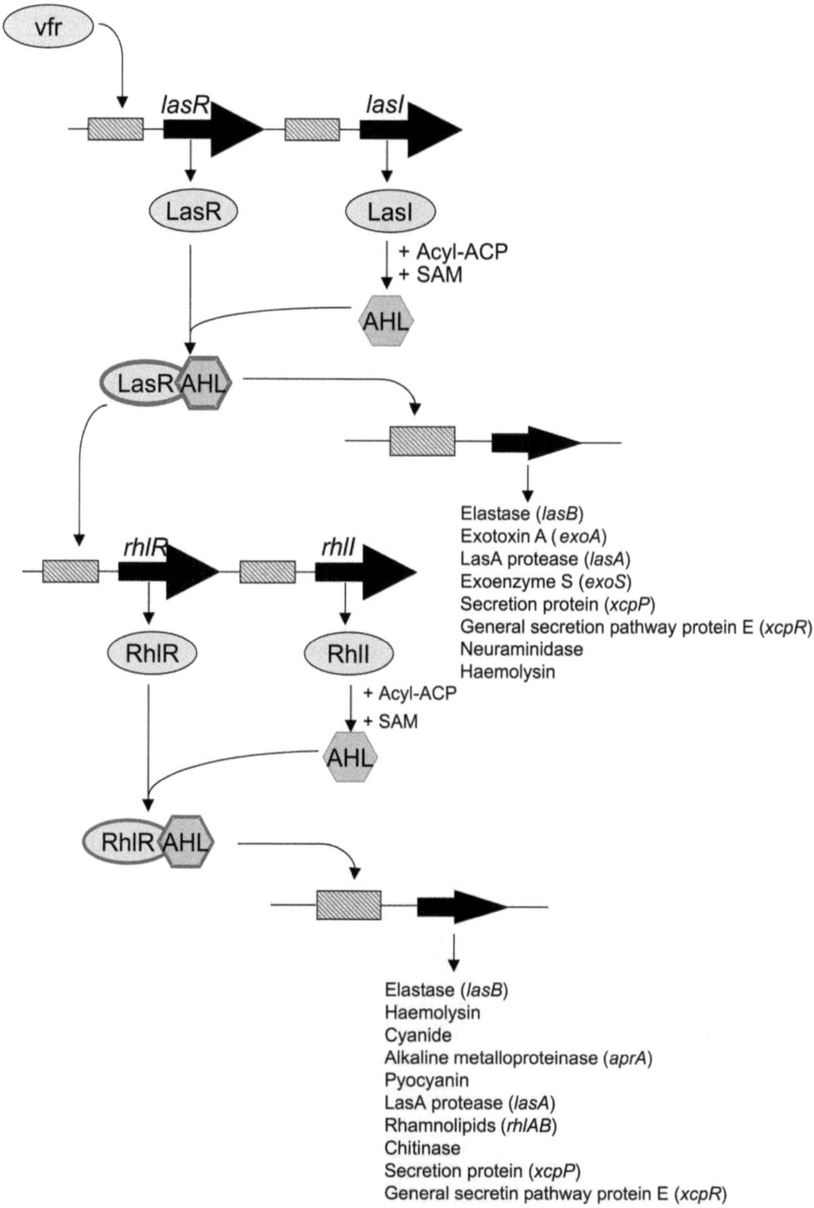

Fig. 5. Overview of quorum sensing (QS) in *Pseudomonas aeruginosa*. Two QS systems (*las* and *rhl*) have been determined in *P. aeruginosa*. Each system is comprised of two genes (*lasR*, *rhlR*) that encode a regulatory protein and an autoinducer synthase (*lasI*, *rhlI*) which catalyses the final step in the synthesis of the acylated homoserine lactone (AHL) signal molecule from an acyl-acyl carrier protein (Acyl-ACP) and *S*-adenosyl methionine (SAM). The regulatory protein and its cognate AHL bind, and the activated complex can then bind to promoter regions (*striped boxes*) of regulated genes. Gene-products and genes regulated by each system are indicated. Some overlap exists in the regulation of genes in *P. aeruginosa* such as *lasB* (elastase), which is regulated by both the *las* and *rhl* QS systems. The LasR-AHL complex is also capable of regulating the expression of both *lasI* and *rhlR*, while the RhlR-AHL complex is capable of regulating expression of *rhlI*

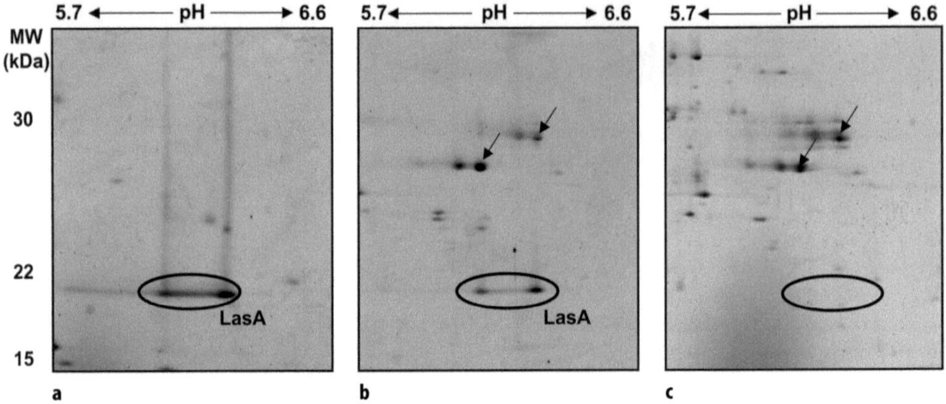

Fig. 6a–c. Influence of QS on extracellular protein expression in *P. aeruginosa*. Extracellular proteins from wildtype and QS mutants were separated on 2-D gels and visualized with SyproRuby. Identical sections of each strain are shown. LasA, regulated by both *las* and *rhl* QS systems is visible: **a** in the wildtype) (*circled*); **b** in the *lasRlasI* mutant, although at a reduced level, as the *rhl* quorum sensing system is still functional; **c** loss of both quorum sensing systems results in complete loss of *lasA* expression. Interestingly, both the double mutant *lasRlasI* (**b**) and the quadruple mutant *lasRlasIrhlRrhlI* (**c**) show expression of proteins not detected in the wildtype (*arrows*)

3.6 Post-Translational Modifications (PTMs)

The role of post-translational modification (PTM) in bacteria is only now starting to become of interest as an important part of gene (protein) function. The function of PTMs include stability, protection from proteases and signal transduction, especially through the functions of two-component regulatory systems that involve the phosphorylation of a response regulator via a histidine kinase to activate a series of further genes and proteins. Potential modifications on bacterial proteins include phosphorylation, glycosylation, methylation, deamidation and biotinylation. Other modifications such as sulfation, hydroxylation and myristoylation do occur naturally on proteins, but at this stage appear specific to eukaryotes. However, as the presence and importance of PTMs on bacterial proteins is only just being realized, no doubt in future investigations other types of PTMs will be found to occur. The effect of PTMs on protein location on a 2-D gel depends on the charge and molecular mass of the modification. Some modifications, in particular phosphorylation and glycosylation, can significantly change the isoelectric point of a protein on a gel relative to its predicted position. Often, heterogeneity in the modification (i.e. the number of modified amino acids within a protein) can result in many spots on a 2-D gel representing one gene. In most cases, modified proteins are recognizable by a 'train' of spots across the pH range of a 2-D gel. A number of proteins in *P. aeruginosa* are known to be phosphorylated. However, very few bacterial proteins in general are known to be glycosylated [66]. Examples include bacterial adhesins such as pili in *Neisseria* sp. and flagellins in *Campylobacter jejuni*. In *P. aeruginosa*, some strains, including

6206, have type A flagellins that are glycosylated, while type B flagellins (e.g. in strain PA01) are not glycosylated [67, 68]. Some heterogeneity in the glycosylation of type A flagellin must occur, at least for strain 6206, as at least three distinct spots corresponding to flagellin have been identified in our studies. Another application of proteomics to observe PTMs examined the alternative sigma factor AlgU [69]. 2-D gels of *P. aeruginosa* proteins were performed following treatment of whole cell lysates with Ser/Thr phosphatase. Phosphatase treated protein maps revealed a change in the number of AlgU isoforms, suggesting that AlgU function is affected by this modification. While many isoforms present on 2-D gels may be artefactual, some major proteins in both the extracellular and outer membrane fraction of *P. aeruginosa* are represented as several spots on 2-D gels, suggesting that PTMs are a common occurrence in these classes of protein. Furthermore, a global 'glycostain' of PA01 OMPs revealed several novel glycosylated proteins including an organic solvent tolerance protein, OmpH8 and Fe-pyochelin receptor protein [39].

4 Differential Protein Expression

One of the main advantages in using a proteomic approach, particularly utilizing 2-DGE, is the visual comparative display of changes in protein expression due to modification of a given set of conditions. Comparative projects can be used to identify gene expression based on different growth conditions or nutrient limitations, growth phase and gene knockout mutagenesis. Such studies can also be used to define the protein differences associated with well-characterized phenotypic differences between strains. A number of studies on *P. aeruginosa* have been comparisons of protein expression based on different growth conditions. For example, Quadroni et al. [45] compared PAO1 proteins expressed in media both with and without sulfate using 2-DGE. Thirteen protein spots were seen for the first time or up-regulated during sulfate starvation and these were characterized by tandem MS/MS or *N*-terminal Edman sequencing. Interestingly, five periplasmic solute-binding proteins of unknown function were induced, suggesting that these are involved in sulfate scavenging. This is an example of how proteomics can reflect the global response of an organism to a given set of environmental conditions.

Differential displays may also be based on comparison of antibiotic sensitive and resistant strains or gene positive and negative strains to view the global effects of phenotype or gene knockout. For example, Jorgensen et al. [70] examined stress tolerance in *P. aeruginosa* through comparison of *rpoS* positive and negative strains. RpoS is an alternative sigma factor of RNA polymerase and regulates the general stress response in Gram negative bacteria. In *E. coli,* RpoS increases during entry to stationary phase and in the presence of certain stresses such as acid, low temperature and hyperosmotic stress. The RpoS homolog in *P. aeruginosa,* not surprisingly, behaves similarly, effecting cell survival under stress, with *rpoS* mutants more sensitive to heat shock, low pH, ethanol and high osmolarity, although the variance was not as pronounced as in *E. coli*. Using 2-DGE, differential expression of four gene products could be detected. Two proteins were down-regulated, one was completely repressed and the fourth was synthesized in the mutant but not the wildtype. However, proteins were separated post-radiolabelling and hence were not identified.

Proteomics has also been utilized to detect strain differences from chronic cystic fibrosis isolates compared to environmental, clinical and initial CF infections [46]. SDS-PAGE revealed that different strains (383 and 2192; non-mucoid and mucoid CF isolates, respectively) expressed very different membrane protein profiles even though they were genetically identical. This result differs from those we have seen with strains PA01 and the cytotoxic isolate 6206, where membrane proteins revealed on 2-D gels are essentially identical even though the strains differ at the genetic level (Nouwens et al., submitted). Whole cell extracts revealed several further protein differences between the strains including a number of 'housekeeping' genes (e.g. RpsS1 and DnaK were unique to 2192). Mucoid strain 383 over-expressed several OMPs including OprF, OMP H1 and potential pathogenicity factors such as chitin-binding protein (*chpD*) and superoxide dismutase (*sodA*). Interestingly, the differences associated with OprF and ChpD between strains were due to differential cleavage, an event we have also detected in our studies of *P. aeruginosa* isolates. Strain characterisations have also been undertaken for several antibiotic resistant and susceptible *P. aeruginosa* isolates [40]. This group confirmed that imipenem resistance could be phenotypically achieved by mutations in OprD that resulted in a lack of functional, membrane-associated OprD. Furthermore, the presence of a beta-lactamase-like protein was also correlated with ceftazidime resistant strains. In total, 100 2-D gel purified proteins were characterized by *N*-terminal Edman sequencing. The technological improvements in proteome analysis and the release of the *P. aeruginosa* genome sequence will mean that further emphasis should be placed on such comparative studies in the future.

4.1 Host-Pathogen Interactions

One of the areas in which proteomics can now be applied is the examination of host-pathogen interactions. Examining protein expression in *P. aeruginosa* grown in isolation is informative, and much can be gained, particularly when differential displays are considered, but expression of proteins that occur strictly in response to host factors may be excluded. For example, the *xcp* (type III) secretion system provides a means by which *P. aeruginosa* can directly inject extracellular factors, including proteins of the exoenzyme S (ExoS) regulon, into host cells [71, 72]. Expression of cytotoxicity factors such as ExoS, Exo, T, and ExoU may be highly specific for the host environment. Few proteomic analyses have been undertaken where two interacting organisms occupy the same niche. However, a good precedent is the analysis of symbiotic relationships between *Rhizobium* and plant species during nodulation [73, 74]. While this is undoubtedly complex due to the dual nature of the interaction and the poor representation of genetic information for these organisms in sequence databases, it does suggest that such studies could be undertaken successfully for host-pathogen interactions such as *P. aeruginosa* in the CF lung. Recently, a similar analysis was undertaken using cDNA microarrays to examine the response of human epithelial cell lines to infection with *P. aeruginosa* [23]; however, the corresponding response in the bacterium was not analysed.

5 Limitations of Proteomics

Proteomics provides a global view of protein expression in response to a given set of conditions, as well as the tools, such as peptide mass mapping via MALDI-TOF MS, for the characterization and subsequent identification of those proteins positively or negatively responding to such conditions. Additionally, proteomic analyses may uncover new ORFs not detected in genomic annotation, as was recently reported for *M. tuberculosis* [75]. Such ORFs tend to encode hypothetical proteins of very low mass that are too small to be accurately predicted via informatics alone. However, limitations still exist in proteomics in terms of the number of proteins that can be resolved using 2-DGE, as well as the physical parameters of proteins that cannot be visualized using this technology. Despite the huge improvements that have been made in 2-DGE and protein sample preparation, several protein types are still poorly represented using this technology. These can be grouped into four separate categories: (1) basic proteins, (2) hydrophobic proteins, (3) high and low mass proteins and (4) low copy number proteins (reviewed in [28]). It is not the objective of this review to analyse these in detail; however, it is important to realize that, as yet, no proteomics approach is capable of resolving an 'entire proteome'. Sub-proteomic type approaches can greatly improve our ability to resolve proteins by 2-DGE and have been applied in a number of studies, yet despite this, a complete proteome has not yet been obtained. Undoubtedly, we will have to compromise by combining complementary technologies such as 2-DGE and multi-dimensional liquid chromatography-mass spectrometry for protein analysis, and rely heavily on the interpretation of transcriptomics experiments to understand events at the regulatory level.

6 Conclusion – Proteomics Beyond the Microbial Genomic Era

As highlighted, there have been only a limited number of proteomic-style projects for *P. aeruginosa*. This may reflect the traditional difficulties associated with protein identification via *N*-terminal Edman sequencing and the poor reproducibility and resolution of pre-IPG 2-D gels. However, the release of the complete genome sequence for this organism, coupled with recent improvements in protein separation, visualization and detection, as well as high sensitivity, high-throughput mass spectrometry approaches, will no doubt result in a range of new proteomic analyses of *P. aeruginosa* in the future. Such analyses will allow us to appreciate a better understanding of *P. aeruginosa* pathogenicity and hopefully lead to better therapies with which to combat this organism.

Acknowledgements. This work has been facilitated by access to the Australian Proteome Analysis Facility, established under the Australian Government Major National Research Facilities program and the Australian Proteome Industry Research and Development (APIRD) grant. ASN is the recipient of an Australian Postgraduate Award and an APIRD Award. The authors wish to thank S. Beatson and C. Whitchurch (The University of Queensland) for the generation of QS mutants and M. Willcox for providing strain 6206. SC wishes to acknowledge Bio-Rad Laboratories and Micromass Ltd for financial and instrumentation support.

7 References

1. Lyczak JB, Cannon CL, Pier GB (2000) Microb Infect 2:1051
2. Bellido F, Hancock R (1993) In: Campa M, Bendinelli M, Friedman H (eds) *Pseudomonas aeruginosa* as an opportunistic pathogen. Plenum Press, New York, p 321
3. Britigan BE, Railsback M, Cox CD (1999) Infect Immun 67:1207
4. Meyer JM (2000) Arch Microbiol 174:135
5. DeVos D, De Chial M, Cochez C, Jansen S, Tummler B, Meyer JM, Cornelis P (2001) Arch Microbiol 175:384
6. Blumer C, Haas D (2000) Microbiology 146:2417
7. Galloway DP (1991) Mol Microbiol 5:2315
8. Iglewski BH, Kabat D (1975) Proc Natl Acad Sci USA 2:2284
9. Hornef MW, Roggenkamp A, Geiger AM, Hogardt M, Jacobi CA, Heesemann J (2000) Microb Pathog 29:329
10. Poole K (2001) Curr Opin Microbiol 4:500
11. Poole K, Tetro K, Zhao Q, Neshat S, Heinrichs DE, Bianco N (1996) Antimicrob Agents Chemother 40:2021
12. Srikumar R, Paul CJ, Poole K (2000) J Bacteriol 182:1410
13. Ziha-Zarifi I, Llanes C, Kohler T, Pechere JC, Plesiat P (1999) Antimicrob Agents Chemother 43:287
14. Hahn HP (1997) Gene 192:99
15. Comolli JC, Hauser AR, Waite L, Whitchurch CB, Mattick JS, Engel JN (1999) Infect Immun 67:3625
16. Poole K, Neshat S, Krebes K, Heinrichs DE (1993) J Bacteriol 175:4597
17. Folschweiller N, Schalk IJ, Celia H, Kieffer B, Abdallah MA, Pattus F (2000) Mol Membr Biol 17:123
18. Stover CK, Pham XQ, Erwin AL, Mizoguchi SD, Warrener P, Hickey MJ, Brinkman FSL, Hurnagle WO, Kowalik DJ, Lagrou M, Garber RL, Goltry L, Tolentino E, Westbrock-Wadman S, Yuan Y, Brody LL, Coulter SN, Folger KR, Kas A, Larbig K, Lim R, Smith K, Spencer D, Wong GK-S, Wu Z, Paulsen IT, Reizer J, Saier MH, Hancock REW, Lory S, Olsen MV (2000) Nature 406:959
19. Croft L, Beatson SA, Whitchurch CB, Huang B, Blakeley RL, Mattick JS (2000) Microbiology 146:2351
20. Brinkmas FSL, Hancock REW, Stover CK (2000) Nature 406:933
21. Spiers AJ, Buckling A, Rainey PB (2000) Microbiology 146:2345
22. Harrington CA, Rosenow C, Retief J (2000) Curr Opin Microbiol 3:285
23. Ichikawa JK, Norris A, Bangera MG, Geiss GK, van't Wout AB, Bumgarner RE, Lory S (2000) Proc Natl Acad Sci USA 97:9659
24. DeLisa MP, Wu CF, Wang L, Valdes JJ, Bentley WE (2001) J Bacteriol 183:5239
25. Schoolnik GK, Voskuil MI, Schnappinger D, Yildiz FH, Meibom K, Dolganov NA, Wilson MA, Chong KH (2001) Methods Enzymol 336:3
26. Gygi SP, Rochon Y, Franza BR, Aebersold R (1999) Mol Cell Biol 19:1720
27. Nouwens AS, Hopwood FG, Traini M, Williams KL, Walsh BJ (1999) In: Charlebois RL (ed) Organization of the prokaryotic genome. American Society for Microbiology, Washington DC, p 331

28. Cordwell SJ, Nouwens AS, Walsh BJ (2001) Proteomics 1:461
29. Washburn MP, Wolters D, Yates JR III (2001) Nature Biotechnol 19:242
30. Gygi SP, Rist B, Aebersold R (2000) Curr Opin Biotechnol 11:396
31. Smolka MB, Zhou H, Purkayastha S, Aebersold R (2001) Anal Biochem 297:25
32. Herbert B, Molloy M, Gooley A, Walsh B, Bryson W, Williams K (1998) Electrophoresis 19:845
33. Chevallet M, Santoni V, Poinas A, Rouquie D, Fuchs A, Kieffer S, Rossignol M, Lunardi J, Garin J, Rabilloud T (1998) Electrophoresis 19:1901
34. Lopez MF, Berggren K, Chernokalskaya E, Lazarev A, Robinson M, Patton WF (2000) Electrophoresis 21:3673
35. Chalmers MJ, Gaskell SJ (2000) Curr Opin Biotechnol 11:384
36. Gygi SP, Aebersold R (2000) Curr Opin Chem Biol 4:489
37. Molloy M, Herbert B, Walsh B, Tyler M, Traini M, Sanchez J-C, Hochstrasser D, Williams K, Gooley A (1998) Electrophoresis 19:837
38. Cordwell SJ, Nouwens AS, Verrills NM, Basseal DJ, Walsh BJ (2000) Electrophoresis 21: 1094
39. Nouwens AS, Cordwell SJ, Larsen MR, Molloy MP, Gillings M, Willcox MDP, Walsh BJ (2000) Electrophoresis 21:3797
40. Michea-Hamzehpour M, Sanchez J-C, Epp S, Paquet N, Hughes G, Hochstrasser D, Pechere J-C (1993) Enzyme Prot 47:1
41. Cowell BA, Willcox MDP, Herbert B, Schneider RP (1999) J Appl Microbiol 86:944
42. Cordwell SJ, Wilkins MR, Cerpa-Poljak A, Gooley AA, Duncan M, Williams KL, Humphery-Smith I (1995) Electrophoresis 16:438
43. Cordwell SJ, Humphery-Smith I (1997) Electrophoresis 18:1410
44. O'Connor CD, Adams P, Alefounder P, Farris M, Kinsella N, Li Y, Payot S, Skipp P (2000) Electrophoresis 21:1178
45. Quadroni M, James P, Dainese-Hatt P, Kertesz MA (1999) Eur J Biochem 266:986
46. Hanna SL, Sherman NE, Kinter MT, Goldberg JB (2000) Microbiology 146:2495
47. Urquhart-Grindlinger BL (2001) PhD Thesis, University of Sydney
48. Masuda N, Sakagawa E, Ohya S (1995) Antimicrob Agents Chemother 39:645
49. Nikaido H, Ocusu H, Ma D, Li X-H (1996) In: Nakazawa T (ed) Molecular biology of pseudomonads. ASM Press, Washington DC, p 353
50. von Specht B, Knapp B, Hungerer K, Lucking C, Schmitt A, Domdey H (1996) J Biotechnol 44:145
51. Mansouri E, Gabelsberger J, Knapp B, Hundt E, Lenz U, Hungerer KD, Gilleland HE Jr, Staczek J, Domdey H, von Specht BU (1999) Infect Immun 67:1461
52. Lee NG, Jung SB, Ahn BY, Kim YH, Kim JJ, Kim DK, Kim IS, Yoon SM, Nam SW, Kim HS, Park WJ (2000) Vaccine 18:1952
53. Molloy MP, Herbert BR, Slade MB, Rabilloud T, Nouwens AS, Williams KL, Gooley AA (2000) Eur J Biochem 267:2871
54. Rabilloud T, Blisnick T, Heller M, Luche S, Aebersold R, Lunardi J, Braun-Breton C (1999) Electrophoresis 20:3603
55. Trias J, Nikaido H (1990) J Biol Chem 265:15,680
56. Ochs MM, Lu CD, Hancock REW, Abdelal AT (1999) J Bacteriol 181:5426
57. Nicas TI, Hancock REW (1983) J Bacteriol 153:281
58. Rawling EG, Brinkman FSL, Hancock REW (1998) J Bacteriol 180:3556
59. Price BM, Galloway DR, Baker NR, Gilleland LB, Staczek J, Gilleland HE Jr (2001) Infect Immun 69:3510
60. De Kievit TR, Iglewski BH (2000) Infect Immun 68:4839
61. Klinger JD, Straus DC, Hilton CB, Bass JA (1978) J. Infect Dis 138:49
62. Cahan R, Axelrad I, Safrin M, Ohman DE, Kessler E (2001) J Biol Chem 276:43,645
63. Whiteley M, Lee KM, Greenberg EP (1999) Proc Natl Acad Sci USA 96:13,904
64. Fuqua C, Greenberg EP (1998) Curr Opin Microbiol 1:183
65. Pesci EC, Iglewski BH (1997) Trends Microbiol 5:132
66. Schaffer C, Graninger M, Messner P (2001) Proteomics 1:248

67. Brimer C, Montie T (1998) J Bacteriol 180:3209
68. Arora SK, Bangera M, Lory S, Ramphal R (2001) Proc Natl Acad Sci USA 98:9342
69. Schurr MJ, Yu H, Martinez-Salazar JM, Hibler NS, Deretic V (1995) Biochem Biophys Res Comm 216:874
70. Jorgensen R, Bally M, Chapon-Herve V, Michel G, Lazdunski A, Williams P, Stewart GSAB (1999) Microbiology 145:835
71. Yahr TL, Mende-Mueller LM, Friese MB, Frank DW (1997) J Bacteriol 179:7165
72. Frank DW (1997) Mol Microbiol 26:621
73. Natera SH, Guerreiro N, Djordjevic MA (2000) Mol Plant Microbe Interact 13:995
74. Morris AC, Djordjevic MA (2001) Electrophoresis 22:586
75. Jungblut PR, Muller EC, Mattow J, Kaufmann SH (2001) Infect Immun 69:5905
76. Goldberg JB, Ohman DE (1987) J Bacteriol 169:1349
77. Kessler E, Safrin M, Abrams WR, Rosenbloom J, Ohman DE (1997) J Biol Chem 272: 9884
78. Vessillier S, Delolme F, Bernillon J, Saulnier J, Wallach J (2001) Eur J Biochem 268:1049
79. Vincent TS, Fraylick JE, McGuffie EM, Olson JC (1999) Mol Microbiol 32:1054
80. Olson JC, Fraylick JE, McGuffie EM, Dolan KM, Yahr TL, Frank DW, Vincent TS (1999) Infect Immun 67:2847
81. Hassan HM, Fridovich I (1980) J Bacteriol 141:156
82. Engels W, Endert J, Kamps MA, van Boven CP (1985) Infect Immun 49:182
83. Goldberg JB, Pler GB (1996) Trends Microbiol 4:490
84. Rocchetta HL, Burrows LL, Lam JS (1999) Microbiol Mol Biol Rev 63:523
85. May TB, Chakrabarty AM (1994) Trends Microbiol 2:151
86. Gacesa P (1998) Microbiology 144:1133
87. Stanislavsky ES, Lam JS (1997) FEMS Microbiol Rev 21:243
88. Feldman M, Bryan R, Rajan S, Scheffler L, Brunnert S, Tang H, Prince A (1998) 66:43
89. Britigan BE, Rasmussen GT, Cox CD (1997) Infect Immun 65:1071
90. Songer JG (1997) Trends Microbiol 5:156

Received: April 2002

Adv Biochem Engin/Biotechnol (2003) 83: 141–176
DOI 10.1007/b11115

Mass Spectrometry – a Key Technology in Proteom Research

Albert Sickmann [1] · Marcus Mreyen [2] · Helmut E. Meyer [3]

[1] Rudolf-Virchow-Zentrum, DFG Forschungszentrum für Experimentelle Biomedizin, Versbacher Strasse 9, Raum 411, 97078 Würzburg, Germany
E-mail: Albert.Sickmann@virchow.uni-wuerzburg.de
[2] PROT@GEN AG, Emil-Figge-Strasse 76A, 44227 Dortmund, Germany
E-mail: marcus.mreyen@protagen.de
[3] Medical Proteom Center, Ruhr University of Bochum, ZKF E.143 44780 Bochum, Germany
E-mail: HelmutE.Meyer@ruhr-uni-bochum.de

The rapid developments in the field of mass spectrometry have transformed it into a key technology in proteome research.

Increased sensitivity in mass spectrometry, as a result of more efficient ionisation techniques and better detectors, has allowed the stepwise reduction of protein quantity for analysis. Protein spots of 2D-PAGE separated samples are now quantitatively sufficient for an unequivocal identification of a protein by mass spectrometry. In addition to protein identification a closer look at posttranslational modifications is now also possible. It is speculated that modifications like phosphorylation or glycosylation exist on every second protein and that they are important for the protein function.

This review highlights the different mass spectrometric methods and gives a brief overview of strategies and methods used to identify modifications.

Keywords. Proteomics, Mass spectrometry, Phosphorylation Glycosylation, Quantification

List of Abbreviations

2,5-DHB	2,5-Dihydroxybenzoic acid
2D	Two dimensional
4-HCCA	α-cyano-4-hydroxycinnamic acid
CID	Collision-induced dissociation
CZE	Capillary-zone-electrophoresis
ESI	Electrospray ionisation
FTICR	Fourier transform ion cyclotron resonance
GC	Gas chromatography
HABA	2-(4′-Hydroxyphenylazo)benzoic acid
HPAEC-PAD	High performance anion exchange chromatography with pulsed amperometric detection
HPLC	High performance liquid chromatography
IAA	Iodoacetamide
LC	Liquid chromatography
m/z	Mass to charge ratio
MALDI	Matrix assisted laser desorption/ionisation
MS	Mass spectrometry
PAGE	Polyacrylamide gel electrophoresis
PNGase-F	Peptide-*N*-glycosidase-F
PSD	Post-source decay
PVDF	Polyvinylidene difluoride

1
Introduction

Sequence analysis of proteins and peptides is not limited to the determination of the primary structure of a protein; furthermore, the analysis of posttranslational modifications is an important task of modern protein chemistry in proteome research. However, the most common application of mass spectrometry in proteome studies is today the identification of proteins derived from 2D gels. In the following sections the developments of mass spectrometry in the last few years, the basic techniques of protein and peptide analysis, data interpretation, analysis of posttranslational modifications and protein quantification are described.

2
Technical Development of Mass Spectrometry

During the last decade, since the introduction of electrospray ionisation [1] and matrix assisted laser desorption ionisation [2] for large biomolecules, these techniques have become the most powerful tools for protein identification and characterization. The common mass spectrometer consists of three units – the ion source, the mass analyser and the ion detection system. The combination of ESI- and MALDI-MS with several different types of mass analysers enables a large number of different mass spectrometers for special purposes. A short overview is given in Table 1.

The most common instruments are ESI ion traps, ESI triple quads, ESI Q-TOF and MALDI-ReTOFs and TOF/TOFs. The combination of ESI- and MALDI ion source with a FTICR mass analyser is still promising but rather expensive for common usage. Besides the technical development of ion sources, mass analysers and ion detection systems, the miniaturized sample preparation protocols lead to enhanced sensitivity.

2.1
Increase of Sensitivity for ESI-MS

The usual ESI ion source operates with solvent flow rates between a few and several hundred microliters per minute depending on the HPLC system connected to the ESI mass spectrometer. The introduction of the nanospray ion source [3] reduces the solvent flow rate to few nanoliters per minute and a more than one hundred-fold increase in sensitivity can be achieved thereby. One major disadvantage of the nanospray technique was the need to analyse several peptides derived from a sample in parallel. Nowadays, the miniaturization of liquid chro-

Table 1. Overview: Mass spectrometer used in protein analysis

Ion source	Mass analyser	Ion detection
Matrix assisted laser desorption/ionisation (MALDI)	Time of flight (TOF and TOF/TOF)	Multi channel plate or electron multiplier
Electrospray ionisation (ESI)	Electromagnetic ion trap (Fourier resistance ion cyclotron) Electric ion trap Triple quadrupole Quadrupole tof (Q-Tof) Quadrupol ion trap (Q-trap) Linear ion trap	Cryer Detector

Table 2. Increase of sensitivity according to the inner diameter of a HPLC column

	Inner diameter [mm]	Relative sensitivity	Flow [μL/min]
Conventional-LC	4.6	≡ 01	> 1000
Analytical-LC	2.0	5	200
Micro-LC	0.5 – 1.0	35	10 – 50
Capillary-LC	0.1 – 0.5	200	1 – 10
Nano-LC	0.05 – 0.1	3500	0.05 – 0.2

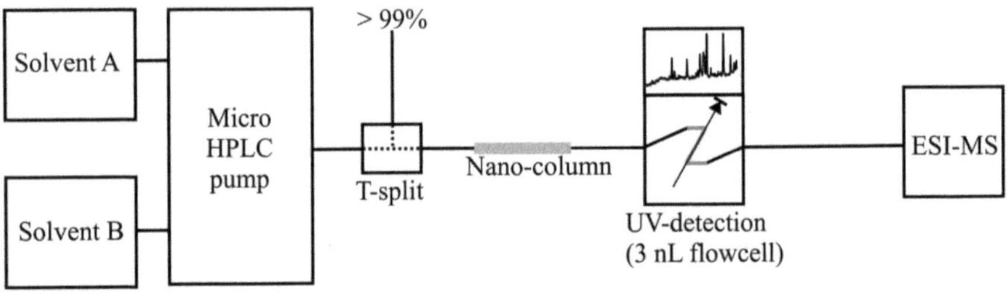

Fig. 1. Instrumental requirements for a nano-HPLC separation. The miniaturization of all capillaries and tubing is strongly recommended. Otherwise resolution and sensitivity of the nano-HPLC decrease dramatically

matography (nano-HPLC, see Table 2) [4] allows a reproducible separation and identification of peptides in combination with nano electrospray ionisation. The technical requirements for a high sensitive nano-LC-ESI-MS/MS system are shown in Fig. 1. The application of 75-μm I.D. nano HPLC columns allows a very sensitive chromatographic separation. However, the applied sample volumes for such HPLC columns must be less than 1 μL. In Fig. 2 an online preconcentration and desalting method which reduces sample volumes between 1 and 100 μL to 0.5 μL is demonstrated.

2.2 Increase of Sensitivity for MALDI-MS

An approach to enhance sensitivity during MALDI-MS analysis is the sample preconcentration with reversed phase column material, e.g. ZipTip (Millipore, Bedford, CA) or a miniaturization of the sample preparation on the MALDI target. The dried droplet sample preparation with α-cyano-4-hydroxycinnamic acid results in a sample spot with a diameter between 1 and 2 mm. An innovative new method was described in 2000 [5]. For this preparation the MALDI target is covered with a hydrophobic surface and small hydrophilic spots with diameters of 200 μm, 400 μm or 600 μm. Samples between 1 and 5 μL are applied onto the hydrophilic spots. During solvent evaporation the droplet starts shrinking but the centre remains on the hydrophilic position and causes a small homogenous sample preparation with a ten- to hundred-fold increase of sensitivity (see Fig. 3).

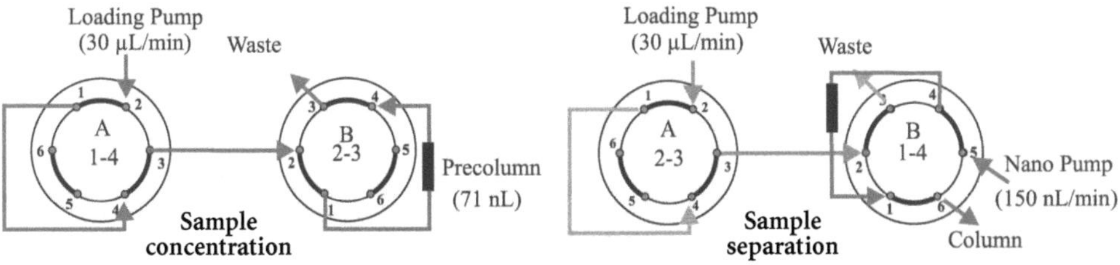

Fig. 2. Online sample preconcentration for nano-HPLC. The online concentration requires two automatic HPLC valves. The settings of valves A and B are demonstrated in the figure. The preconcentration is done with a 0.3 mm × 1 mm C18 column with a flow of 50 µL/min for 10 min. The sample volume can easily be reduced from 100 µL to 0.5 µL (volume of the precolumn and capillaries) during the preconcentration step. The following peptide separation is done with a 75 µm I.D. column with a flow rate of 150 nl/min

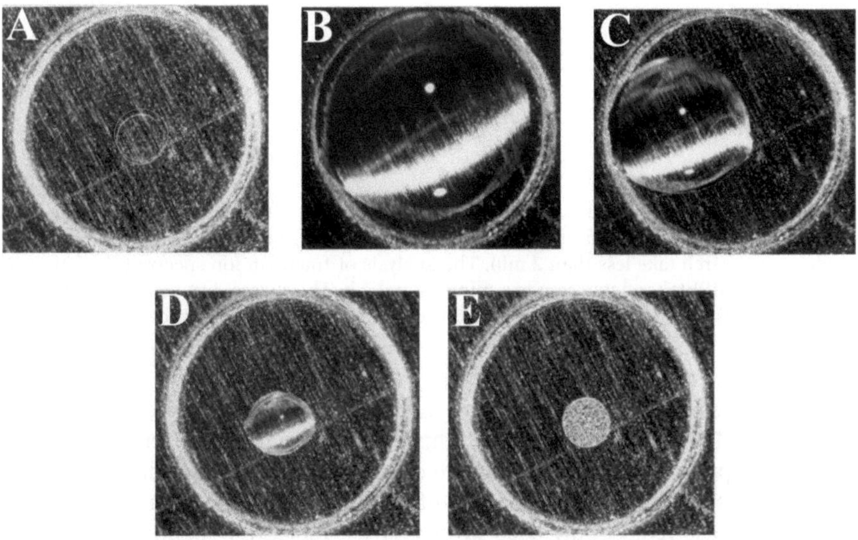

Fig. 3A–E. Concentration of peptide samples using the Anchor Chip technology. Hydrophilic anchor on a hydrophobic surface. Diameter 400 µm. Peptide sample in matrix solution is applied. The droplet start shrinking and is fixed on the anchor position. The droplet is nearly dried. Final micro preparation for MALDI-MS

3 Sequence Analysis of Proteins and Peptides by Mass Spectrometry

3.1 Identification of Proteins Using Peptide Mass Fingerprint

Peptide mass fingerprints are the fastest method for the identification of proteins recovered from 2D-PAGE. After gel electrophoresis the protein spots are punched out, washed and digested with a specific protease (see Fig. 4). The generated pep-

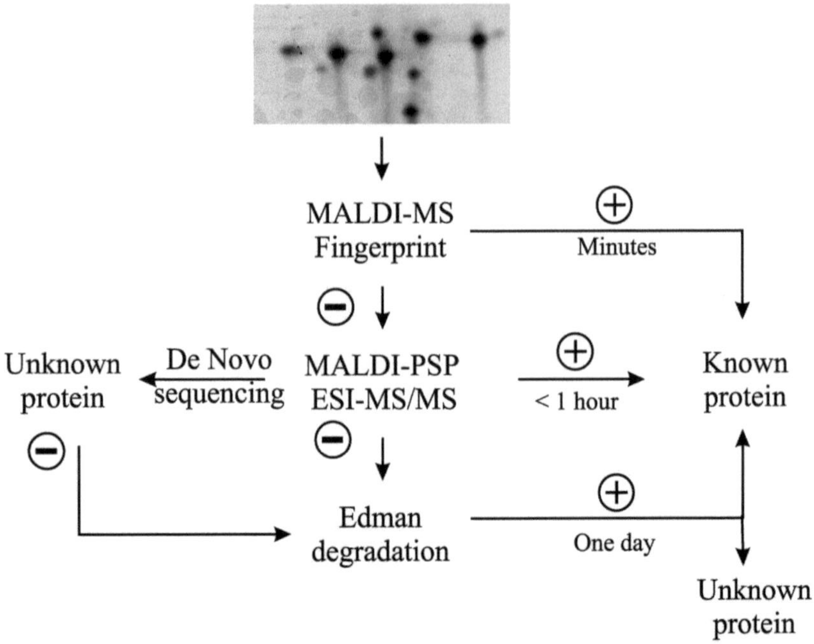

Fig. 4. Overview: possible strategies for protein analysis after in gel digestion. The fastest method for protein identification is a MALDI mass fingerprint. The measurement time and the database search take less than 2 min. The analysis of fragment ion spectra (MALDI-MS/MS or LC-MS/MS) takes between a few minutes and 2 h. The slowest but database-independent method is the Edman degradation which takes about 60 min per sequenced amino acid

tides are eluted in a small volume and subjected to mass spectrometry, e.g. MALDI-MS. The basic principle of this technique is the comparison of the measured peptide masses with calculated peptide masses from database entries. Every protein results in a unique set of peptide masses after cleavage with a specific protease (commonly used proteases see Table 3). Depending on the mass accuracy and mass resolution of the used instrument only few peptide masses are required for a reliable protein identification. An example for a peptide mass fingerprint is given in Fig. 5.

3.2
Identification Using Peptide Fragmentation

An alternative method for the identification of proteins and peptides is the fragmentation of an isolated peptide ion by post source decay after MALDI [6, 7] or collision induced dissociation [8] after ESI. Both techniques lead to a statistical fragmentation at the peptide bound of a peptide. The generated ions are called b-ions when the *N*-terminus is incorporated and y-ions when the *C*-terminus is a part of the fragment ions [9]. The mass difference of two following b- or y-ions correspond to an amino acid mass (see Fig. 6) and, there-

Table 3. Overview: Proteases used in protein analysis

Endopeptidase	Type	Specificity	pH range	Inhibitors
Chymotrypsin	Serine	Y, F, W	1.5–8.5	Aprotinin, DFP, PMSF
Trypsin	Serine	R, K	7.5–9.0	TLCK, DFP, PMSF
Glu C	Serine	D, E	7.5–8.5	DFP
Lys C	Serine	R	7.5–8.5	DFGP, Aprotinin, Leupeptin
Arg C	Cysteine	R	7.5–8.5	EDTA, Citrate
Asp N	Metallo	D (N-terminal)	6.0–8.0	EDTA
Elastase	Serine	A, V, I, L, G	8.0–9.0	DFP, a1-Antitrypsin, PMFS
Pepsin	Acidic	F, M, L, W	2.0–4.0	Pepstatin
Subtilisin	Serine	Nearly all	7.0–11.0	Phenol, DFP, PMSF
Thermolysine	Metallo	Hydrophobic AA	7.0–9.0	EDTA
Papaine	Cysteine	R, K, G, H, Y	7.0–9.0	IAA, TLCK, TPCK
Proteinase K	Serine	Hydrophobic AA	7.0	IAA
Thrombin	Serine	R	7.5	DFP, TLCK, PMSF
Factor X	Serine	I-E-G-R	8.3	DFP, PMSF

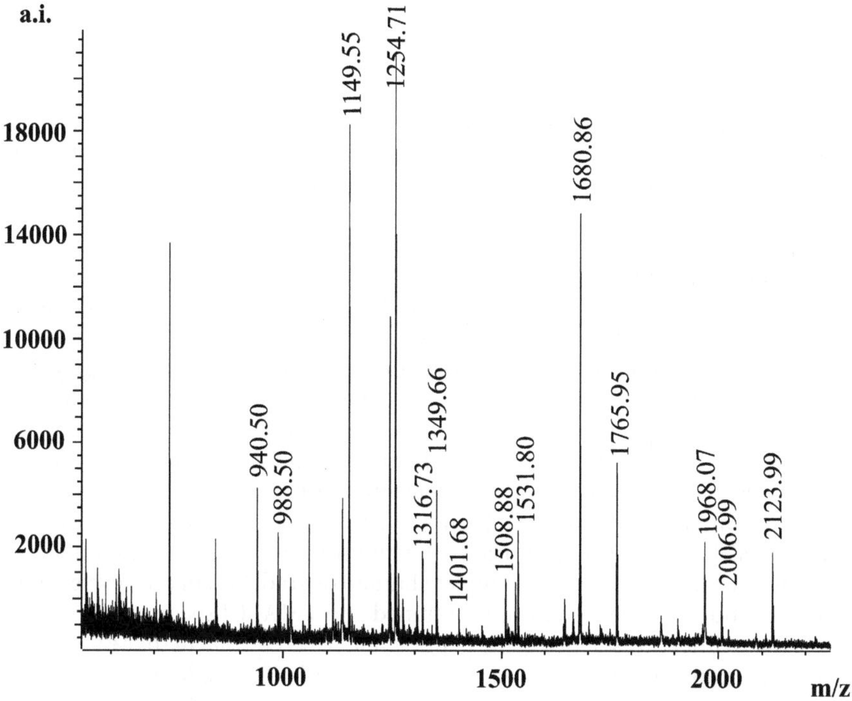

Fig. 5. Peptide mass fingerprint spectrum of a yeast protein. The database interpretation of this mass fingerprint spectrum with the algorithm ProFound results in endonuclease SceI 75 kDa subunit (gi|171463) from *Saccharamyces cerevisiae* (YOR F!) YJR045 C

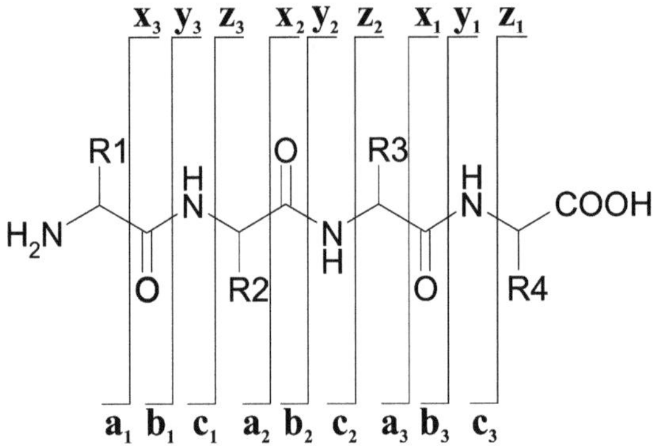

Fig. 6. Fragmentation of peptides during MALDI-PSD and ESI-MS/MS. The shown nomenclature was introduced by Biemann [9]. Under MALDI conditions the most prominent ions are a-, b- and y-ions and after ESI the b- and y-ions

fore, a PSD or CID fragment ion spectrum contains the sequence information of a peptide.

3.3 Identification of Proteins Derived from Organisms with Unknown Genome

If the genome of an organism is unknown the interpretation of the MALDI fingerprint spectra and fragment ion spectra cannot be done against a database and therefore some other approaches are necessary for a successful identification of proteins.

De novo sequencing of peptides with mass spectrometry is based on the manual interpretation of a-, b- and y-ion series in a mass spectrum. This kind of data interpretation is very time-consuming and the results always need to be checked against the fragment ion spectrum of synthetic peptides.

A promising approach is the incorporation of ^{18}O into peptide fragments by using endoproteases [10]. This results in *C*-terminal fragment ions with a mass shift of 2 or 4 Da. i.e. using a mixture of $H_2{}^{16}O/H_2{}^{18}O$ generates ion doublets. Together with accurate mass determination of peptide fragments this method increases the precision of sequence determination [11].

A combination of classical Edman chemistry and mass spectrometry was described in 1993 [12]. The usage of a mixture of phenylisothiocyanate and phenylisocyanate results in incomplete Edman degradation in each cycle. The reaction products of each cycle are combined and a MALDI spectrum is recorded. The mass differences between the peptide masses is equivalent to the mass of the cleaved amino acid in each step. The usage of volatile coupling reagents allows very fast cleavage of the *N*-terminal amino acid and sensitivities in the femtomol range are reached [13]. However, this method has some limitations. The two amino acids leucine and isoleucine cannot be distinguished because of their same mass. Small differences in mass as with glutamine and lysine requires an instrument with high mass accuracy and resolution. Modern MALDI mass spectrometers can be applied to this technique in a mass range up to 5000 Da.

A third method is the coupling of MALDI mass spectrometry and Edman degradation.

Another strategy to identify unknown proteins is the usage of an offline micro-HPLC MALDI-MS system combined with Edman degradation. As an example of this the sequence analysis of an up to now unknown protein is demonstrated. The protein was digested overnight with trypsin; the peptides are extracted and separated by preparative micro-HPLC using a 180 μm I.D. × 250 mm C18 column (Fig. 7). The column eluate is automatically spotted onto a MALDI target (20% of each fraction; see Fig. 8) and additional PVDF membrane stripes (80% of each fraction). The MALDI instrument is used to determine the number of peptides in each fraction (Fig. 9) and to record MALDI-PSD spectra of the peptides. The corresponding fraction on the PVDF stripe is selected for Edman degradation and the sequence results are compared with the sequence information derived from MALDI-PSD. The combination of Edman analysis and mass spectrometry sequence information allows one to deduce more than only one peptide sequence from a single HPLC fraction especially when two or more peptides coelute from the column.

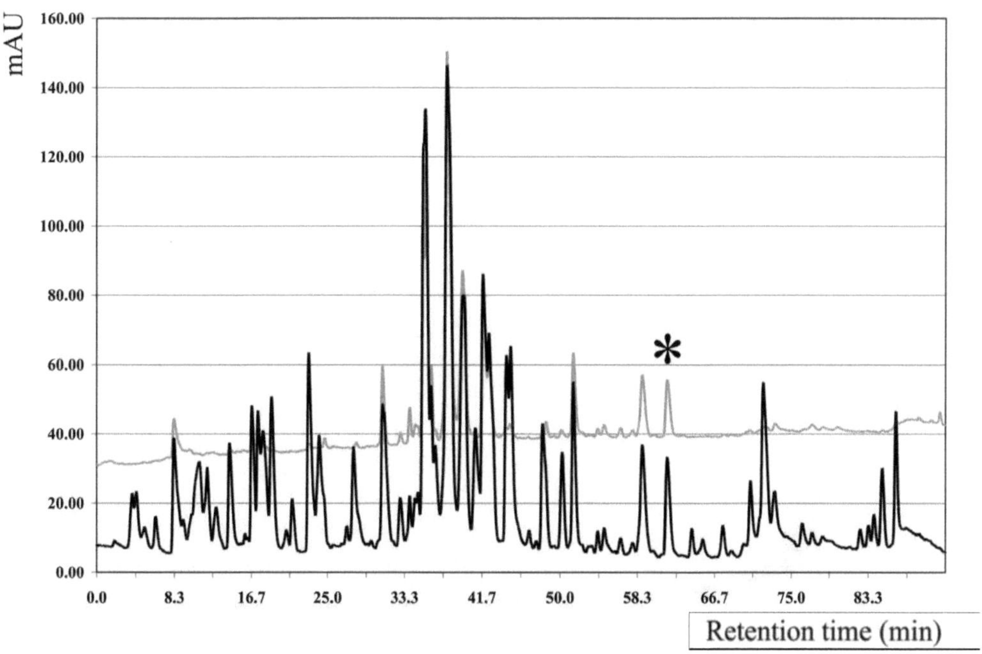

Fig. 7. Micro-HPLC separation of tryptic peptides. The digestion of the unknown protein was done with trypsin overnight. The separation is done with a common TFA/acetonitrile/water gradient system. UV traces are recorded at 215 nm (*black*, peptide backbone) and 295 nm (*grey*, tryptophane side chain). The mass spectrum in Fig. 9 results from the fraction marked with *

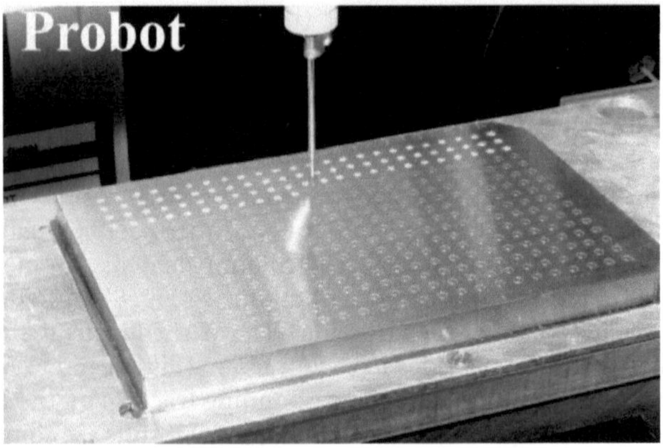

Fig. 8. Automatic spotting of HPLC fractions. The column eluate is automatically spotted onto a MALDI target. The robot needle remains for 60 s on a single position and moves then to the following position. The dual collect mode allows a parallel spotting onto PVDF membrane and MALDI target. In this case the robot needle remains for 2 s on the MALDI target followed by 8 s on the PVDF membrane. This alternate spotting is repeated six times (60 s) and than the robot needle moves to the next position. Because of this alternate spotting of one HPLC fraction the same peptides are collected onto the PVDF membrane and the MALDI target

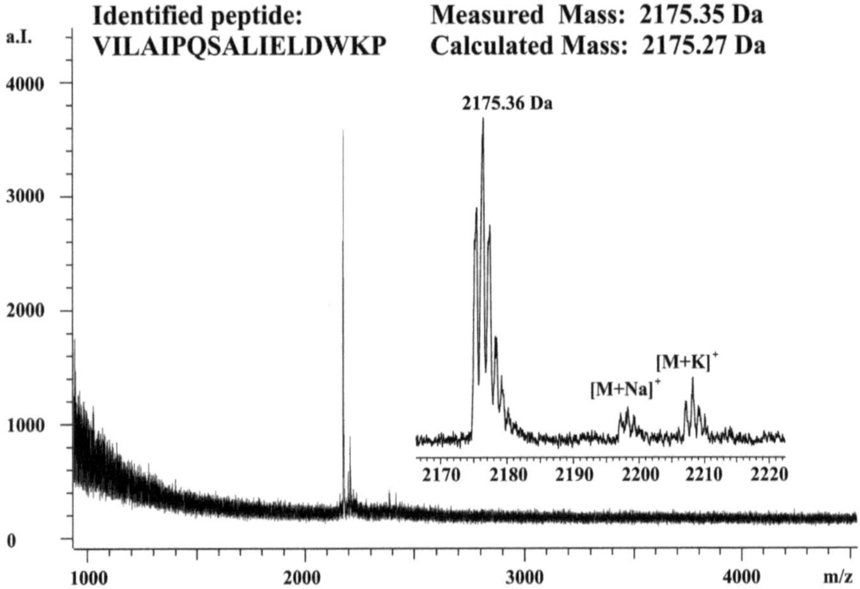

Fig. 9. MALDI-MS spectrum of a micro HPLC fraction. This fraction was subjected to Edman degradation and the received sequence VILAIPQSALIELDWKP fits to the measured mass of 2175.35 Da

Table 4. Overview: Web accessible algorithms and databases

Search engines	Web address
Mascot	http://www.matrixscience.com/search_form_select.html
MSFit	http://prospector.ucsf.edu
dbEST	http://ncbi.nlm.nih.gov/blast/
NCBInr	http://ncbi.nlm.nih.gov/blast/
SwissProt	http://us.expasy.ch/sprot
csm/Dictyostelium discoideum	http://www.csm.biol.tsukuba.ac.jp/
OWL	http://www.bioinf.man.ac.uk/dbbrowser/owl

3.4 Data Interpretation Using the Internet

Besides sample preparation the data interpretation is one of the fundamentals for a successful analysis of protein and peptide samples. Several different algorithms are available from the world wide web and allows data interpretation via the internet. All programs are also commercially available for a more confidential data interpretation in an intranet. An overview of the used programs is given in Table 4.

4 Analysing Posttranslationally Modified Proteins

Sequence analysis is not only limited to the elucidation of the primary structure of a protein; in addition, attached posttranslational modifications are included in the analysis. These modifications are important for protein interaction in cell recognition, signal-transduction or protein localization. In the following paragraphs some approaches to analyse the most common modifications like glycosylation and phosphorylation are described.

4.1 Glycosylation

Glycoproteins are proteins with different carbohydrates covalently bound at distinct amino acids within the protein backbone. These proteins modified with this common and highly diverse co- and posttranslational modification are found in a broad range of organisms. Glycoproteins are found inside the cell, both in cytoplasm and the subcellular organelles, in cell membranes and in extracellular fluids. In fact, most proteins are glycoproteins, which include all kinds of biologically active substances, such as enzymes, receptors, antibodies, hormones and cytokines, as well as structural proteins such as collagen.

Increasing the sensitivity of the methods used to detect and analyse accurately glycosylation has revealed the diversity in both structures and functions of the carbohydrate components of proteins. The glycan structures of glycoproteins are often heterogeneous and are built up of a manageable number of monosaccharides (Fig. 10).

Variation at a single site of glycosylation is termed microheterogeneity and diversity between sites of glycosylation is termed macroheterogeneity. This combined heterogeneity results in many discrete subsets or glycoforms of glycoproteins and sometimes can be observed by two-dimensional electrophoresis (2D-PAGE) as typical "trains" of protein-spots which separate on the basis of different isoelectric point (pI) and/or molecular mass.

Diversity has been observed for carbohydrates attached to a single protein produced by different tissues or organisms and even in one cell type glycosylation may vary depending on the cell cycle, state of differentiation and development. In the era of proteomics where protein patterns are compared, these modification on proteins, which changes the protein physicochemical properties, should not be ignored.

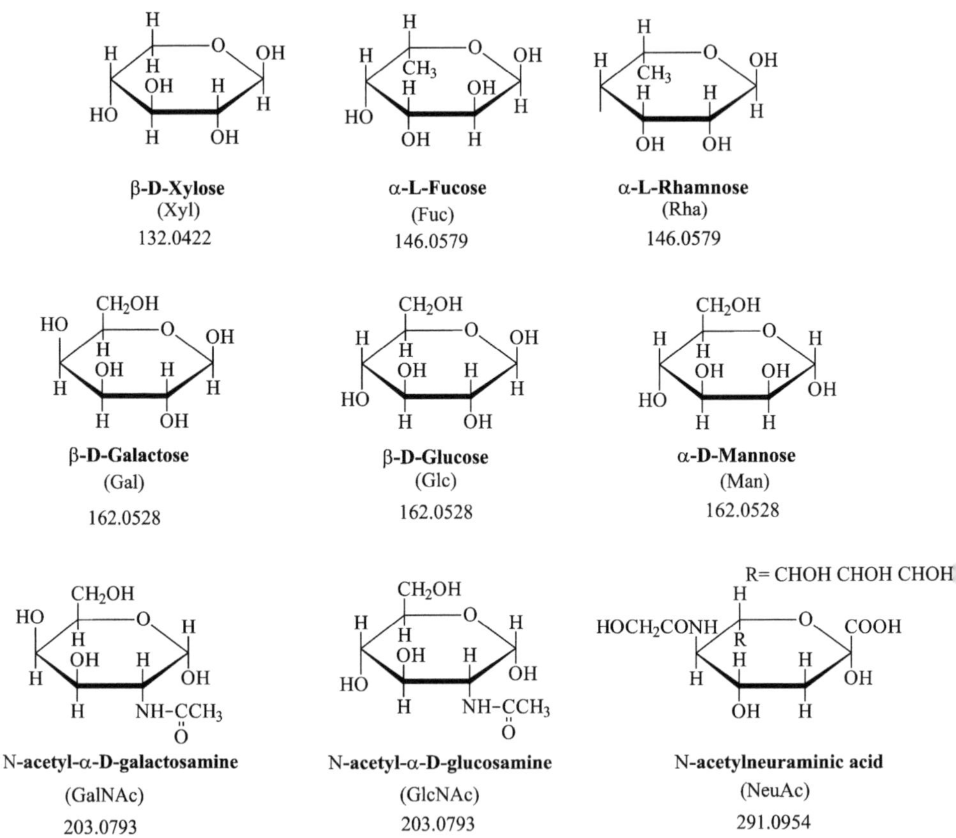

Fig. 10. Common monosaccharide constituents of glycoproteins α/β, anomeric configuration; D/L, optical configuration; monosaccharide abbreviation in brackets; listed value, monoisotopic mass of underivatised monosaccharide residue

It has to be noticed that complex carbohydrates present a significant challenge: the sugar type, anomeric configuration, the linkage position between monomeric units, branching, the sequence of sugar epimers, and the ring size all contribute to the enormous structural diversity of this class of modification. Other biological substitutions (e.g. phosphorylation and sulfation) complicate things further.

The methods generally used to elucidate glycan structures are high performance anion exchange chromatography with pulsed amperometric detection (HPAEC-PAD) and capillary-zone-electrophoresis (CZE) for composition and oligosaccharide profiling analysis. Gas chromatography-mass spectrometry (GC-MS), a classical method, is used for composition analysis and elucidation of the glycan linkage via methylation/acetylation analysis. Earlier studies were done by FAB-MS on derivatised and non-derivatised glycan structures but are now more and more displaced by more sensitive methods like MALDI-MS, ESI-MS and nanospray-MS. Fragmentation spectra of glycans, obtained by collision induced dissociation (CID) or post source decay (PSD), can give further information about linkages between the monosaccharides.

A method which allows a full glycan characterisation is nuclear magnetic resonance spectroscopy (NMR) but it has its limitation in the large quantities needed for the analysis.

4.1.1
Forms of Glycosylation

A recently published review [14] describes the different types of glycosylation. The most common two forms are called *N*-linked and *O*-linked glycosylation depending on the attachment sites at the protein. A further type of glycosylation mentioned is the glycosylphosphatidyl-inositol (GPI)-anchor where the carboxy terminus of a protein is attached via ethanolamine to the carbohydrate structure containing inositol, which in turn is connected to a hydrophobic lipid membrane anchor. Following the three uncommon glycosylation types are listed:

1. Phosphoglycosylation where a carbohydrate linkage through a phosphodiester bond to a hydroxy amino acid exist, such as Ser or Thr
2. *C*-Mannosylation where the indole ring of tryptophan (Trp) has a mannose residue attached
3. Ribosylation where the linkage of the 3-hydroxyl group of ribose to glutamic acid (Glu), asparagines (Asp), arginine (Arg) or cysteine (Cys) is observed

Up to now most of the protein glycosylation is found on eucaryotic proteins although it now becomes more and more evident that glycosylation in procaryotes combines a much greater diversity of glycan composition, linkage units, and glycosylation sequences on polypeptides [15].

4.1.2
N-Linked Glycosylation

N-Linked glycosylation in eukaryotes is the best analysed form of glycosylation so far. The reducing terminal sugar is covalently bound to the amide group of the

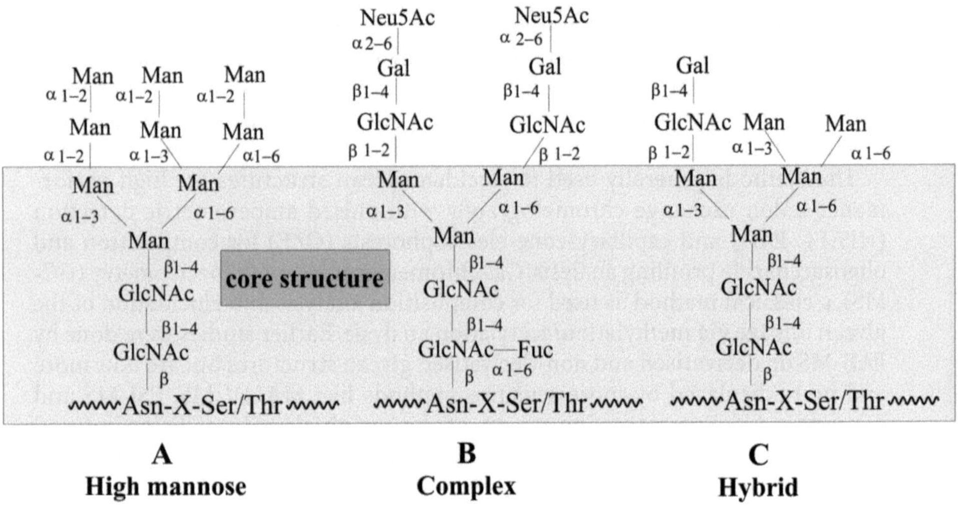

Fig. 11. Representative structures of different types of *N*-linked glycans. Monosaccharide abbreviation see Fig. 10

amino acid asparagine. A core structure of the glycan consisting of $GlcNAc_2Man_5$ is further modified by addition of four mannose and three glucose residues, resulting in the structure $GlcNAc_2Man_9Glc_3$. Finally, this glycan is transferred en bloc to the growing polypeptide at the amino acid asparagin residue in the consensus tripeptide sequence Asn-X-Ser/Thr (X$\neq$Pro) via an *N*-acetylglucosamine (GlcNAc) residue.

After transfer the glycan undergoes reconstruction, being trimmed and rebuilt to form three structural types of *N*-linked oligosaccharides (Fig. 11). These structures are called A) high mannose-type, B) complex-type and C) hybrid type, but may possess a broad range of structural variation.

In the last decade, several new carbohydrate structures linked to asparagines have also been discovered mainly in bacteria glycoproteins and all glycans are linked via the asparagine of the tripeptide Asn-X-Ser/Thr consensus sequence mentioned above [14]. Common structural principles such as the chitiobiose core and antennae of eucaryotic *N*-glycans, however, are not present in prokaryotic glycoproteins. Gram-positive bacteria and archaea show here, as a reducing terminal sugar, for example Glc, GalNAc or Rha attached to Asn [15].

4.1.3
O-Linked Glycosylation

O-linked glycosylation is the second type of glycosylation commonly found in glycoproteins (see Table 5). The reducing terminal sugar is bound to the hydroxyl group of an amino acid, commonly serine (Ser) or threonine (Thr) and less frequently tyrosine (Tyr), hydroxylysine (Hly) or hydroxyproline (Hpr).

It has no define core glycan structure and shows a wide variety of reducing terminal sugar residues.

Table 5. Carbohydrate-hydroxy amino acid linking group

Reducing terminal linkage	Amino acid	Occurrence
Ara*f*[a]	Hyp	Plants
α-Fuc	Ser, Thr	Animals
α-Gal	Hyp	Plants, eubacteria
α-Gal	Ser	Plants, eubacteria
β-Gal	Hyl	Animals (collagen only)
β-Gal	Tyr	Eubacteria
α-GalNAc	Ser, Thr	Eukaryotes
α-Glc	Tyr	Animals (glycogen only)
β-Glc	Ser, Thr	Eubacteria, animals
α-GlcNAc	Ser, Thr	Protozoa
β-GlcNAc	Ser, Thr	Animals
α-Man	Ser, Thr	Yeast, animals
β-Xyl	Ser	Animals

Monosaccharide abbreviation – see Fig. 10.
[a] Ara*f*, Arabinofuranose.

The size may vary from a single unit up to elongated oligosaccharides with more than 100 residues in length, as found on some surface proteins. O-Glycosylation has been observed at single sites or accumulated in clusters in proteins, like in mucins, in contrast to eucaryotic *N*-linked glycosylation with its mostly well separated glycosylation sites.

4.2 Mass Spectrometric Approaches to Glycoprotein Analysis

The identification of proteins out of 1D or 2D gels is generally done by MALDI-MS fingerprint analysis or fragmentation spectra received by MALDI-PSD or ESI-MS/MS analysis. This identification in combination with database information led to a first hint about a potential glycoprotein. Proteins separated by 2D-PAGE and blotted can easily be tested for glycosylation by using commercially available carbohydrate specific staining methods.

For a full characterisation of protein glycosylation the following questions have to be answered:

1. How many individual molecular species (glycoforms) of the protein exist?
1. Which sites of the protein are glycosylated and to what extent are these sites modified?
3. What do the attached glycan structures look like and what is their amount at each individual site?

Mass spectrometry does not allow one to answer all these questions, but with its high sensitivity it has at least opened the door in the Proteomics era for a closer look at this form of modification [16–20].

4.2.1
Investigation of the Intact Glycoprotein

The investigation of intact glycoproteins is still very difficult and has its limitation for all known ionisation techniques, which results in the low number of published MS data for intact glycoproteins. MALDI, ESI and nano-electrospray can be used for ionisation and TOF detectors to obtain a good resolution. Before analysis the sample needs a good clean up of salts and buffers. In MALDI analysis, e.g. the glycoprotein can be adsorbed onto a gold target or onto nitrocellulose layer on the target, followed by a washing step with water to remove the contaminants. Afterwards, the matrix is directly added to the sample and used for analysis [21].

However, MALDI is limited to smaller proteins compared to ESI. With MALDI different glycoforms of up to 35-kDa proteins gave a sufficient resolution, but only when delayed extraction was used [22]. It has been reported that HABA is a superior matrix for large glycoproteins compared to sinapinic acid, 2,5-DHB or 4-HCCA [23]. The signals of glycoproteins are often weaker compared to their unglycosylated form as shown by the changed signal intensity of an isolated membrane protein after the carbohydrate moieties were removed [24]. Although MALDI-TOF shows generally insufficient resolution for large glycoproteins it was used in linear mode to determine the mass of the protomeric subunit of an intact *S*-layer glycoprotein of *T. thermosaccharolyticum*. After depositing the sample on top of the matrix α-cyano-4-hydroxycinnamic acid, the mass was determined to be 75.621 Da, a result which deviated from that obtained by SDS-PAGE by approximately 7.4 kDa [25].

With ESI a good resolution of protein glycoforms could be obtained up to 80 kDa three glycosylation variants of human serum transferrin [26].

Often the large glycoparts cover the surface of the protein and prevent the access to protonation sites, resulting in a much lower average charge state which may be well beyond the *m/z* range of, e.g. quadrupole instruments. A further complication is their tendency to bind salt cations and anions which then will deteriorate signal quality [27].

For all mass spectrometric methods it is better to have glycoproteins with a single glycosylation site and only little microheterogeneity to minimize the production of broad and unresolved peaks. In many mammalian proteins acidic sugar units are found, such as sialic acid, complicating things further because they are both acid labile and tend to cleave easily during sample preparation and ionisation. Although the different isoforms of glycoproteins can sometimes be separated in individual spots by 2D-PAGE, their intact analysis by mass spectrometry afterwards can be hampered by the difficulties to extract large proteins from the gel.

So far nano-electrospray MS, with its extremely small initial-droplet size and increased ionisation efficiency, has given the best spectra of intact glycoproteins with low quantities [27].

4.2.2
Investigation of Glycopeptides

One advantage in investigation of glycopeptides is the reduced mass size which results in better resolution and furthermore the still preserved glycosylation site information which is often lost after releasing the attached glycan.

For investigation of glycopeptides the glycoprotein is digested with proteases in gel or on PVDF blots after separation. Digestion is normally achieved with trypsin followed by reduction and alkylation although other peptidases and cyanogen bromide have also been used. An alternative approach to reduce the protein molecular weight is the digest with unspecific enzymes like pronase which leaves only single amino acids or a short peptide attached to the glycan (Fig. 12).

Different strategies and the help of specific enzymatic tools make it possible to extract information from the modified peptide and the attached glycan out of the generated MS data.

Peptide/glycopeptide mixture can be analysed with different MS techniques mentioned above. Glycopeptides are however difficult to detect because their signals are often suppressed by those of nonglycosylated peptides in a mixture.

Nanospray ionisation shows some advantages in sensitivity compared to "nano-flow" "micro-" ESI ion sources; however, it has, like the MALDI-MS technique, the disadvantage of a missing inline preseparation step for glycopeptides. One way to circumvent this problem is the use of lectins like concavalin A immobilised to paramagnetic beads as a first step in order to isolate glycopeptides [28].

An HPLC-ESI-MS system compared to nanospray-MS has the advantage of an on-line peptide separation before MS analysis, for example on an inline C8 column, where glycosylated peptides generally elute earlier than their non-glycosylated forms due to its distinctive hydrophilic character.

A "daughter-ion-scan" usable on some MS instruments is also helpful for the identification of glycosylated peptides, because of the appearance of diagnostic sugar oxonium ions such as *m/z* 204 (HexNAc), 292 (sialic acid), 163 (hexose) observed by in-source collision-induced dissociation (CID).

Often the source-orifice potential is switched from high at the beginning of each scan to low voltage for the remainder of the scan to identify the modified peptides in the total ion chromatogram.

Once a glycopeptide has been identified the use of tandem mass spectrometry (ESI-MS/MS) is able to elucidate the peptide sequence if there is sufficient material. When the sequence of the determined protein is known the mass of the attached oligosaccharide can be deduced by the mass difference.

Sometimes MALDI-MS fingerprint spectra of glycosylated proteins show a number of unmatched mass signals with low intensity in the area between 3000 and 4500 *m/z* units, resulting from glycopeptides. The signal size can occasionally be improved by enzymatic removal of terminal sialic acid residues from glycans [29].

A comparison of glycosylated with non glycosylated peptides after releasing of the sugars can give further information, because weak glycopeptide

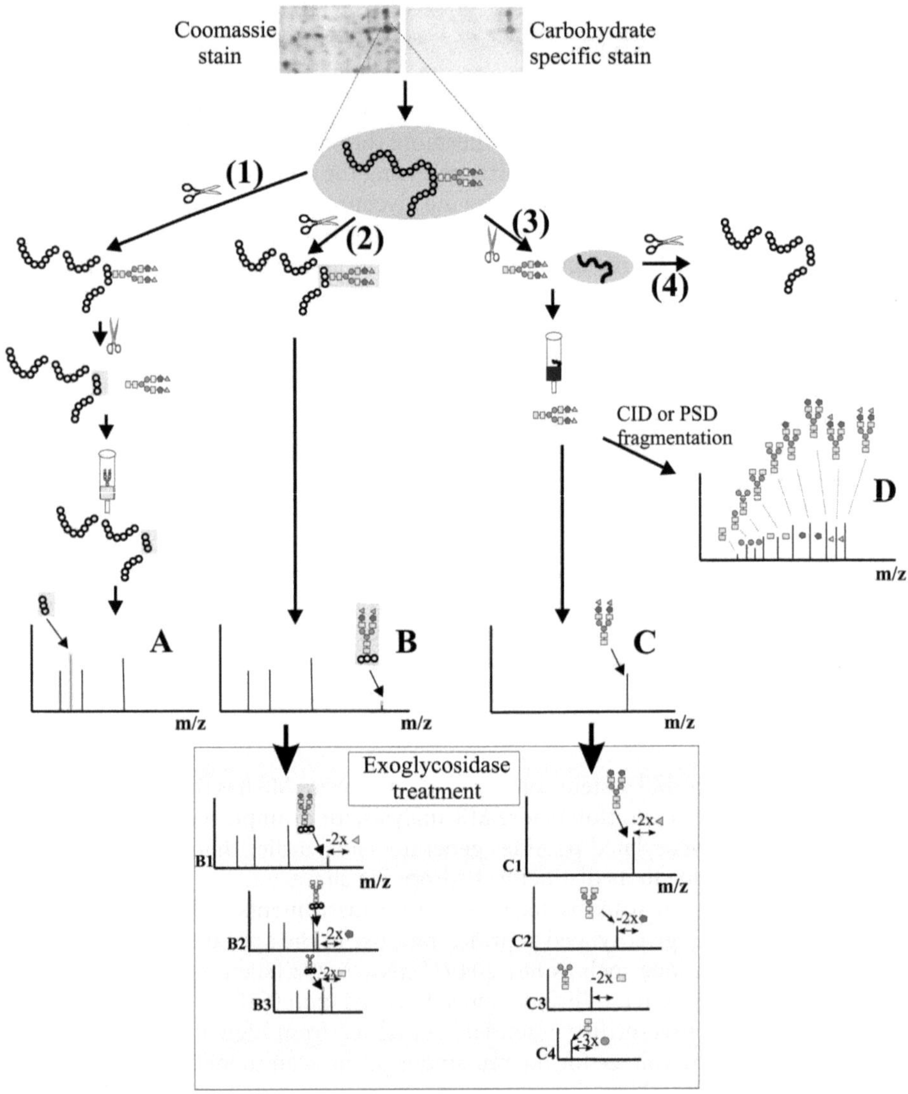

Fig. 12. Example of a flow scheme for glycoprotein characterisation by mass spectrometry. Identification of glycosylated proteins after in gel separation, blotting and staining with carbohydrate specific stains. **(1)** Cut out the protein spot, *horizontal scissors* digest with protease, *vertical scissors* enzymatical protein deglycosylation (e.g. PNGase F), separation of peptides (C8) and analysis by MS. **(2)** Cut out the protein spot, digest with protease and analysis by MS (comparison of spectrum A and spectrum B allows estimation of attached glycan). **(3)** Cut out the protein spot, protein deglycosylation (enzymatically or chemically), extraction and clean up of the released glycans (e.g. on graphitized carbon) prior MS analysis. Samples received by way 2 and 3 may be further used for sequential exoglycosidase treatments to get information about linkage and composition of containing sugars. CID or PSD fragmentation, spectrum D, might gives furthermore an overview about the monosaccharide composition of the removed glycan. **(4)** Releasing glycoprotein glycans in gel without destroying the protein backbone allows afterwards protein identification by MALDI-MS fingerprint analysis

signals disappear whereas new corresponding peptide signals are coming up stronger.

There are some software tools available like GlycoMod (http://expasy.cbr.nrc.ca/tools/glycomod/) which help to calculate, with the knowledge of the right protein sequence and therefore the peptide masses, the possibly linked glycan structures [30].

To release the carbohydrate moieties from glycoproteins you have to differentiate between a chemical and enzymatic cleavages, which function without destroying the peptide structure and this can be done stepwise. Mass spectrometry in combination with structural specific glycosidases is currently the most powerful technique for providing sequence, branching and linkage data from low glycoprotein quantities. The pool of commercially available exo- and endo-glycosidase enzymes has grown in the last years; however, there are still more specific endo-glycosidases available for eucaryotic *N*-linked and less for the large variety of *O*-linked glycosylation (Tables 6 and 7).

Commonly used endoglycosidase enzymes for *N*-linked glycosylation are PNGase-F, PNGase A, EndoF and EndoH which cleave with different specificity on the $Man_3GlcNAc_2$ core structure. In the case of PNGase-F the intact glycan is released under conversion of the modified asparagine to aspartic acid ($m/z+1$) at the *N*-linked consensus sequence site of the protein. With sufficient glycopeptide material the sample can be divided into several fractions and used for exoglycosidases treatment (Fig. 13). This allows the stepwise removal of monosaccharides (e.g. sialic acid, galactose, mannose and *N*-acetylhexosaminidase), and is carried out with single enzymes or with enzyme combination for glycan sequencing [31].

On the basis of the generally available low amounts of material, protocols have been developed which now allow the sequence analysis of glycopeptide glycans or released glycans directly on the MALDI plate [32].

Table 6. Commonly used endoglycosidases for glycan release list adapted from GLYKO catalogue 2001/2002

Enzyme	Specificity	Source
PNGase F	*N*-linked oligosaccharides	*F. meningosepticum*
Endoglycosidase H	*N*-linked hybrid and high mannose oligosaccharides	*Streptomyces plicatus*
Endoglycosidase-F_1	*N*-linked hybrid and high mannose oligosaccharides	*F. meningosepticum*
Endoglycosidase-F_2	*N*-linked biantennary and high mannose oligosaccharides	*F. meningosepticum*
Endoglycosidase-F_3	*N*-linked biantennary and triantennary oligosaccharides	*F. meningosepticum*
O-Glycanase	*O*-linked Galβ1–3GalNAc	*Streptococcus pneumoniae*
Endo-β-galactosidase	Internal β1–4 galactose linkages	*Bacteroides fragilis*

Table 7. Commonly used exoglycosidases for oligosaccharide sequencing list adapted from GLYKO catalogue 2001/2002

Enzyme	Specificity	Source
Sialidase	NeuAc/NeuGc α2-↓-3,6,8,9	*Arthrobacter ureafaciens*
Sialidase	NeuAc/NeuGc α2-↓-3,6,8	*Clostridium perfringens*
Sialidase	NeuAc/NeuGc α2-↓-3,6,8	*Vibrio cholerae*
Sialidase	NeuAc/NeuGc α2-↓-3 > 6 ≫ 8,9	*Salmonella typhimurium*
Sialidase	NeuAc/NeuGc α2-↓-3,8	Newcastle Disease Virus
Sialidase	NeuAc/NeuGc α2-↓-3	*Streptococcus pneumoniae*
β-Galactosidase	Galβ1-↓-4,6,3	Jack bean meal
β-Galactosidase	Galβ1-↓-3,4,6	Bovine testes
β-Galactosidase	Galβ1-↓-3,6	*Xanthomonas manihotis*
β-Galactosidase	Galβ1-↓-4	*Streptococcus pneumoniae*
α-Galactosidase	Galα1-↓-3,4,6	Green coffee beans
β-*N*-Acetylhexosaminidase	GlcNAc/GalNAc β1-↓-2,3,4,6	Jack bean meal
β-*N*-Acetylhexosaminidase	GlcNAc β1-↓-2,3,4,6	*Streptococcus pneumoniae*
α-*N*-Acetylhexosaminidase	GalNAc α1-↓-3	Chicken liver
α-Mannosidase	Man α1-↓-2,3,6	Jack bean meal
α-Mannosidase	Man α1-↓-2	*Aspergillus saitoi*
α-Mannosidase	Man α1-↓-6	*Xanthomonus manihotis*
β-Mannosidase	Man β1-↓-4	*Helix pomatia*
α-Fucosidase	Fucα1-↓-3,4	Almond meal
α-Fucosidase III	Fucα1-↓-3,4	*Xanthomonus manihotis*
α-Fucosidase II	Fucα1-↓-2	*Xanthomonus manihotis*
α-Fucosidase	Fucα1-↓-2,3,4,6	Bovine kidney

Chemical cleavages can also been used to release *N*-linked and *O*-linked glycans; however, they often have the disadvantage of complete destruction of all peptide bonds and therefore the loss of information relating to the glycan attachment site.

Furthermore, these cleavages can generally not be used to release monosaccharides sequentially from glycopeptides.

To determine only the sites of *O*-glycosylation, also in the high density mucin type glycosylation, one-step deglycosylation/alkylaminylation procedures are described for an in-gel sample and liquid sample, respectively [33, 34]. The glycosylated sites are directly modified by an alkylamine label after releasing the sugars, by which a following digestion of the protein is promoted. Attached alkylamine residues make the peptide less polar than the corresponding glycopeptide and, moreover, introduces a positive charge, which also enhances the yield of primary ions detectable in the positive mode of MALDI-MS and ESI-MS.

4.2.3 *Investigation of Glycans*

In order to release glycans from the glycoprotein you can choose between enzymatic cleavage as described above or a chemical cleavage. Most methods are

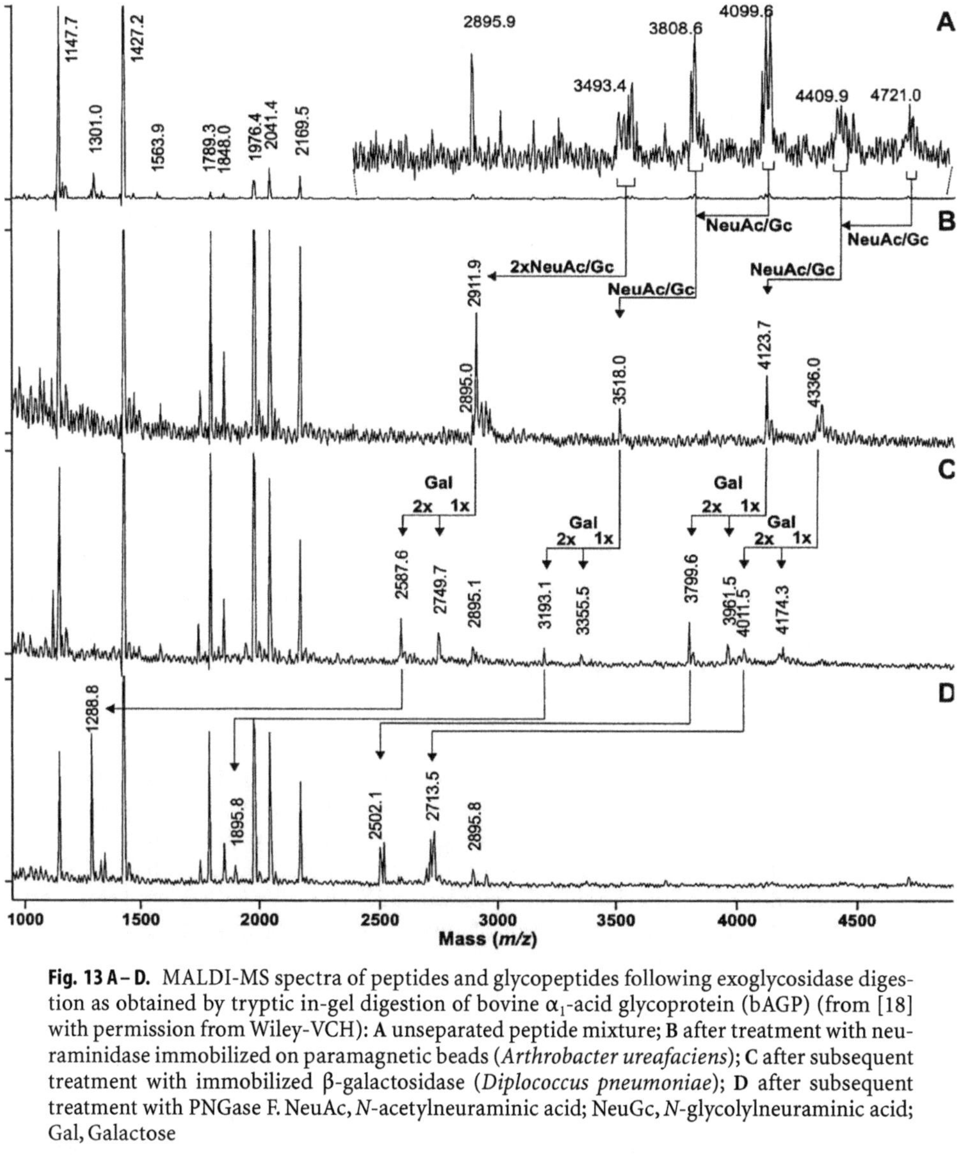

Fig. 13 A – D. MALDI-MS spectra of peptides and glycopeptides following exoglycosidase digestion as obtained by tryptic in-gel digestion of bovine α_1-acid glycoprotein (bAGP) (from [18] with permission from Wiley-VCH): **A** unseparated peptide mixture; **B** after treatment with neuraminidase immobilized on paramagnetic beads (*Arthrobacter ureafaciens*); **C** after subsequent treatment with immobilized β-galactosidase (*Diplococcus pneumoniae*); **D** after subsequent treatment with PNGase F. NeuAc, *N*-acetylneuraminic acid; NeuGc, *N*-glycolylneuraminic acid; Gal, Galactose

accessible for glycoproteins in-gel, on PVDF membrane, or liquid sample. A chemical cleavage with hydrazine is used to remove both glycosylation types. *O*-linked sugars are specifically released at 60 °C, whereas 95 °C is needed to release the *N*-linked glycans. Hydrazine also causes a couple of modifications on some sugars themselves, which then cannot been retraced by mass spectrometry. An improved method is described by Cooper et al. for hydrazinolysis of *O*-linked glycans in combination with triethylamine at lower temperature [35]. The more common way to release *O*-linked glycans is β-elimination with an alkali like NaOH.

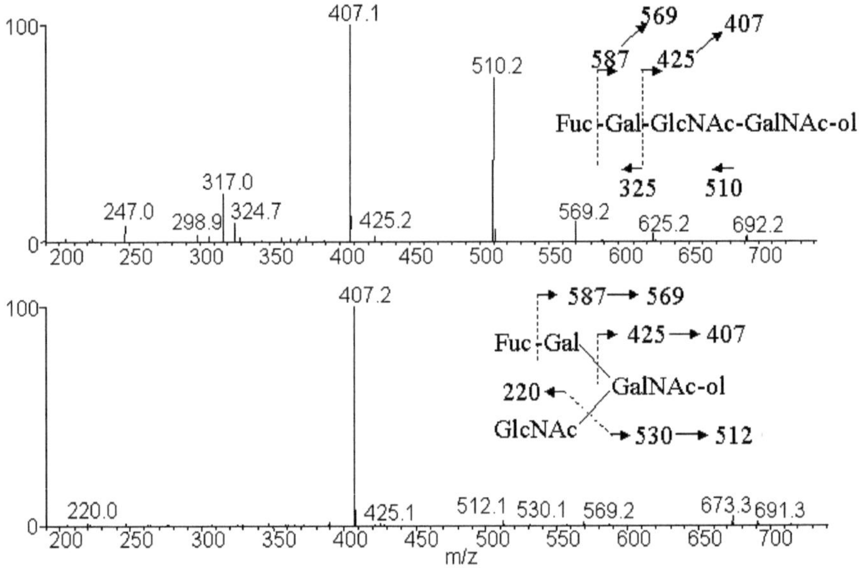

Fig. 14. Fragmentation spectrum of a reduced *O*-linked mucin oligosaccharide of composition dHex1Hex1HexNAc$_2$ (parent mass 936 *m/z*). Electrospray MS/MS spectra obtained on a ThermoFinnigan LCQ Deca after separation on a porous graphitised carbon capillary column (data supplied by Proteome Systems Ltd, Sydney, Australia)

However, reducing terminal sugars are unstable at this high pH and undergo a so-called "peeling reaction". To prevent this degradation generally sodium borohydrate is added to convert the reducing terminal sugar to the alditol.

Carbohydrates usually show also sodium or potassium adduct ions in MALDI and ESI due to the present of salt (*m/z* Na^+, +22 Da; K^+, +38 Da, etc.) (Fig. 14). The increased sensitivity of state of the art mass spectrometer, allows more and more the analysis of native glycans.

Naven and Harvey have examined the influence of glycan structure on ion abundance measured by MALDI-MS on a time-of-flight and on a magnetic sector instrument with 2,5-dihydroxybenzoic acid as the matrix. The signal strength of the $[M+Na]^+$ ion from 25 underivatised oligosaccharides (linear, and both *O*- and *N*-linked oligosaccharides from glycoproteins) with masses greater than about 1000 Da exhibited similar signal strengths, irrespective of structure, when examined on the time-of-flight instrument. Oligosaccharides with masses below 1000 Da displayed a progressive reduction in signal intensity with decreasing molecular weight. A comparable study performed on a magnetic sector instrument revealed that all oligosaccharides studied produced signals of equivalent intensity and that no reduction in signal strength occurred with the smaller sugars [36].

To increase sensitivity derivatisation of the oligosaccharides can be helpful. Derivatives are most commonly added to the reducing terminus of the glycans by a reductive amination reaction with an aromatic amine.

An increase in sensitivity can also be observed by derivatising with Girard's T reagent which introduces a cationic site [37]. An increase in sensitivity of 50- to 1000-fold has been achieved by ligating the released oligosaccharide to a synthetic peptide as peptides are usually ionised efficiently by protonation of their basic groups [38].

The condition of released glycan samples is of critical importance for obtaining high quality MS spectra; therefore many methods for removing salts, buffers, and the rest of derivative reagent and peptides or protein have been developed. It can be differentiated between desalting steps on membranes or on short columns of ion-exchange or hydrophobic resins. The first includes drop dialysis on membranes with a reasonable molecular cut-off [39], or cleaning on Nafion-117 membranes which furthermore remove the peptide or protein part [40]. Clean up on columns is generally done in micro-columns composed of Gel loader tips and some microliters of resin which can be used in different varieties to remove a broad range of contaminants. Mixed-bed ion-exchange columns, however, contain the risk of the loss of charged sugars. A robust method investigated by Packer is the clean up on graphitised carbon as solid phase extraction material [41]. It was used for purification of oligosaccharides or their derivatives with a high recovery rate from solution containing one or more of the following contaminants: salts (including salts of hydroxide, acetate, phosphate), monosaccharides, detergents (SDS and Triton X-100), protein (including enzymes) and reagents for the release of the oligosaccharides from glycoconjugates (such as hydrazine and sodium borohydrate). The sample was applied on a 0.5-mL off-line carbon column, as well as on an on-line Hypercarb HPLC (70 µm, 10 × 4.6 mm) cartridge of a mass spectrometer, followed by an intensive water wash. Neutral oligosaccharides were eluted with 25% solution of acetonitrile in water, and acidic glycans were recovered by further addition of 0.05% TFA to this solution.

A good choice for the analysis of carbohydrates with MALDI-MS is the matrix 2,5-DHB and a mixture with several other compounds, which further increase the sensitivity and resolution. A broad overview of the analysis of carbohydrates by MALDI with all the different used matrices is given by Harvey [42].

As with peptides the different mass spectrometric methods allow the generation of PSD and CID spectra of glycans, which may give further information about sequence, branching and the linkage between the sugars. The observed types of fragmentation are the loss of the adduct, glycosydic cleavages that result from the breaking of a bond linking two sugar rings, and cross-ring cleavages that involve the breaking of two bonds. The detailed nomenclature for describing the fragmentation of carbohydrates is introduced by Domon and Costello [43].

A long used but less sensitive method for the elucidation of glycan linkages is GC-MS analysis. Glycans are generally reduced to prevent "peeling" before the free hydroxyl groups are methylated in a first step. The hydrolysis with TFA breaks down the complex structure in single monosaccharides with unmodified hydroxyl groups at the linkage positions, which are then acetylated in a second derivatising step. This complex mixture is separated and analysed by GC-MS. MS spectra resulting from fragmentation are used for a comparison with database entries of mass spectra of derivatised standard samples.

4.3 Phosphorylation

Phosphorylation of amino acid residues in proteins plays a major role in biological systems. Often phosphorylation acts as a molecular switch controlling the protein activity in different pathways as in metabolism, signal transduction, cell division etc. Therefore, identification of phosphoamino acids in proteins is an important task in protein analysis. In this section an overview of phosphopeptide analysis is presented. Detailed protocols and examples for each step of analysis are described in the given references. Four different types of phosphoamino acid residues are known:

1. *O*-phosphates (*O*-phosphomonoesters) are formed by phosphorylation of hydroxyamino acids such as serine, threonine or tyrosine. The phosphorylation of hydroxyproline or -lysine is as yet unknown.
2. *N*-phosphates (phosphoamidates) are generated by phosphorylation of the amino groups in arginine, lysine or histidine.
3. Acylphosphates (phosphate anhydrides) are produced by the phosphorylation of aspartic or glutamic acid.
4. *S*-phosphates (*S*-phosphothioesters) are formed by phosphorylation of cysteine.

The chemical stability of phosphorylated amino acids is shown in Table 8.

All *O*-phosphates are stable under acidic conditions, in the presence of hydroxylamine, and pyridine. The *N*-phosphates except for phosphoarginine are stable under alkaline condition. All acylphosphates are reactive phosphoamino acids and are labile in acid, alkali, hydroxylamine and pyridine.

Phosphocysteine is moderately stable under all tested conditions.

Before the development of mass spectrometry for large biomolecules (ESI- and MALDI-MS) the only available method for the localisation of phosphoamino

Table 8. Stability of phosphoamino acids

Nature of phosphoamino acid	Stability in			
	Acid	Alkali	Hydroxylamine	Pyridine
O-Phosphates				
Phosphoserine	+	–	+	+
Phosphothreonine	+	±	+	+
Phosphotyrosine	+	+	+	+
N-Phosphates				
Phosphoarginine	–	–	–	–
Phosphohistidine	–	+	–	–
Phospholysine	–	+	–	–
Acylphosphates				
Phosphoaspartate	–	–	–	–
Phosphoglutamate	–	–	–	–
S-Phosphates				
Phosphocysteine	(+)	+	+	+

acids in peptide sequences was Edman degradation. Through the conversion of *P*-Ser to *S*-ethyl cysteine [44] or *P*-Thr to β-methyl-*S*-ethyl cysteine [45] a positive evidence for *P*-Ser and *P*-Thr is possible. *P*-Tyr is stable during Edman degradation but nearly insoluble in the conventional transfer solvents used in the Edman sequenator, yielding a gap in the sequence course. Applying solid phase sequencing to this problem provides a solution [46]. Since the introduction of mass spectrometry, the combination of Edman degradation and mass spectrometry is a powerful tool for localisation of phosphoamino acid residues in protein sequences.

The major problem in analysing phosphorylation with mass spectrometry is the signal suppression of phosphopeptides in a mixture. Therefore a high resolution separation of the mixture before analysis or during analysis is essential for a successful identification of phosphorylated peptides. Due to the fact that radioactive labelling of the incorporated phosphate is easily done with [^{32}P], the method of choice is a separation of the [^{32}P]-phosphopeptide in a peptide mixture followed by high sensitive MALDI-MS analysis of the radioactive fractions. The three *O*-phosphates show different behaviour during MALDI mass spectrometry and fragmentation analysis. However, phosphothreonine and phosphoserine lose their phosphate group after fragmentation (see Fig. 15). Phosphotyrosine is more stable and therefore the phosphate group remains on this amino acid. The generation of dehydroalanine from phosphoserine and α-amino-butyric acid from phosphothreonine can usually be observed with these two phosphoamino acids. This behaviour therefore sometimes makes it impossible to localize the phosphoamino acid by ESI-MS/MS or MALDI-PSD experiments.

If a [^{32}P] labelling is not possible, the analytical strategy has to be changed to LC-MS/MS analysis, e.g. nano-HPLC coupled to an ion trap-mass spectrometer. In a single LC-MS/MS analysis more than 1000 different peptides can be identi-

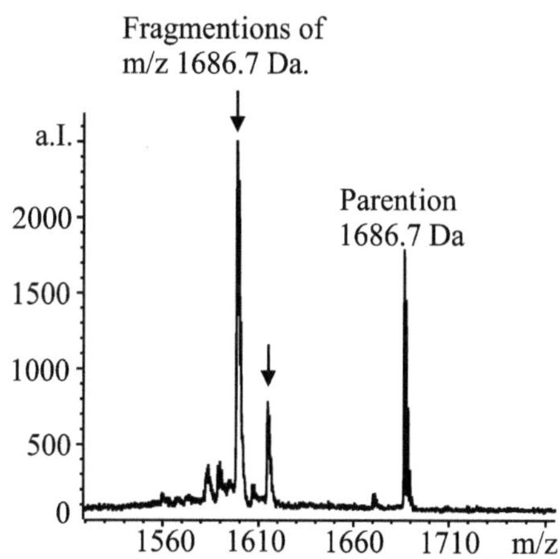

Fig. 15. Typical phosphate cleavage of phosphoserine and phosphothreonine containing peptides. The signals *marked with an arrow* result from a cleavage of phosphate and phosphoric acid

fied within 1 h. With this method a fast identification of posttranslationally modified proteins is possible.

4.3.1
Localisation of O-Phosphates

As shown in Table 8, *O*-phosphates are stable under acidic condition. Therefore the separation of a proteolytic digest containing such phosphopeptides is possible with the common acetonitrile/water/TFA gradient system. The usage of inert HPLC pumps and columns is strongly recommended, because all kinds of phosphopeptides adsorb irreversibly on etched iron surfaces. Using MALDI-PSD for structure determination, the MALDI-target should be inert (e.g. goldsurface). When using steel targets the 'sandwich' preparation technique might be helpful.

To record a MALDI-PSD spectrum of a pure phosphopeptide, commonly less than 50 fmol are required. In mixtures with unphosphorylated peptides (e.g. after digestion) the ionisation rate of the phosphopeptide is impaired and more substance is necessary. Therefore, a separation of the mixture or an enrichment of the phosphopeptide is important.

The localisation of the *O*-phosphates is demonstrated in the following example: localization of phosphotyrosine in human Gab-1.

This protein was over-expressed as a GST fusion protein in *E. coli*. Tyrosin phosphorylation with the insulin receptor kinase takes part in the presence of [^{32}P]γ-ATP. The phosphorylated protein was digested with trypsin overnight and the tryptic peptides were subjected to two-dimensional chromatography. In the first dimension anion exchange chromatography was done and fractions of 0.5 volume were collected (Fig. 16). Fractions containing radioactivity are selected for a μHPLC separation. The rechromatography of these fractions is shown in Fig. 17. Fractions containing radioactivity are marked with an asterisk.

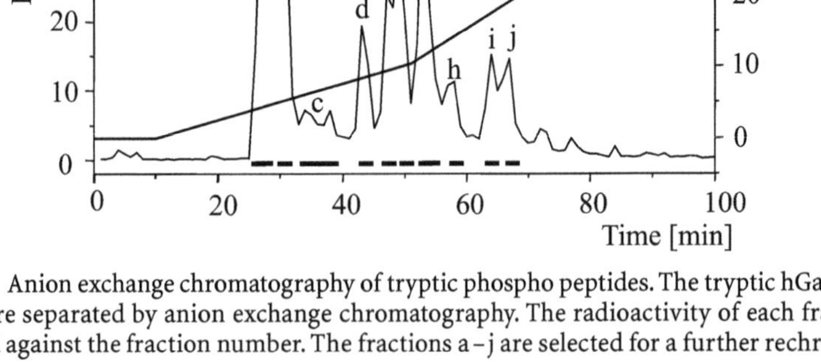

Fig. 16. Anion exchange chromatography of tryptic phospho peptides. The tryptic hGab-1 peptides are separated by anion exchange chromatography. The radioactivity of each fraction is plotted against the fraction number. The fractions a–j are selected for a further rechromatography

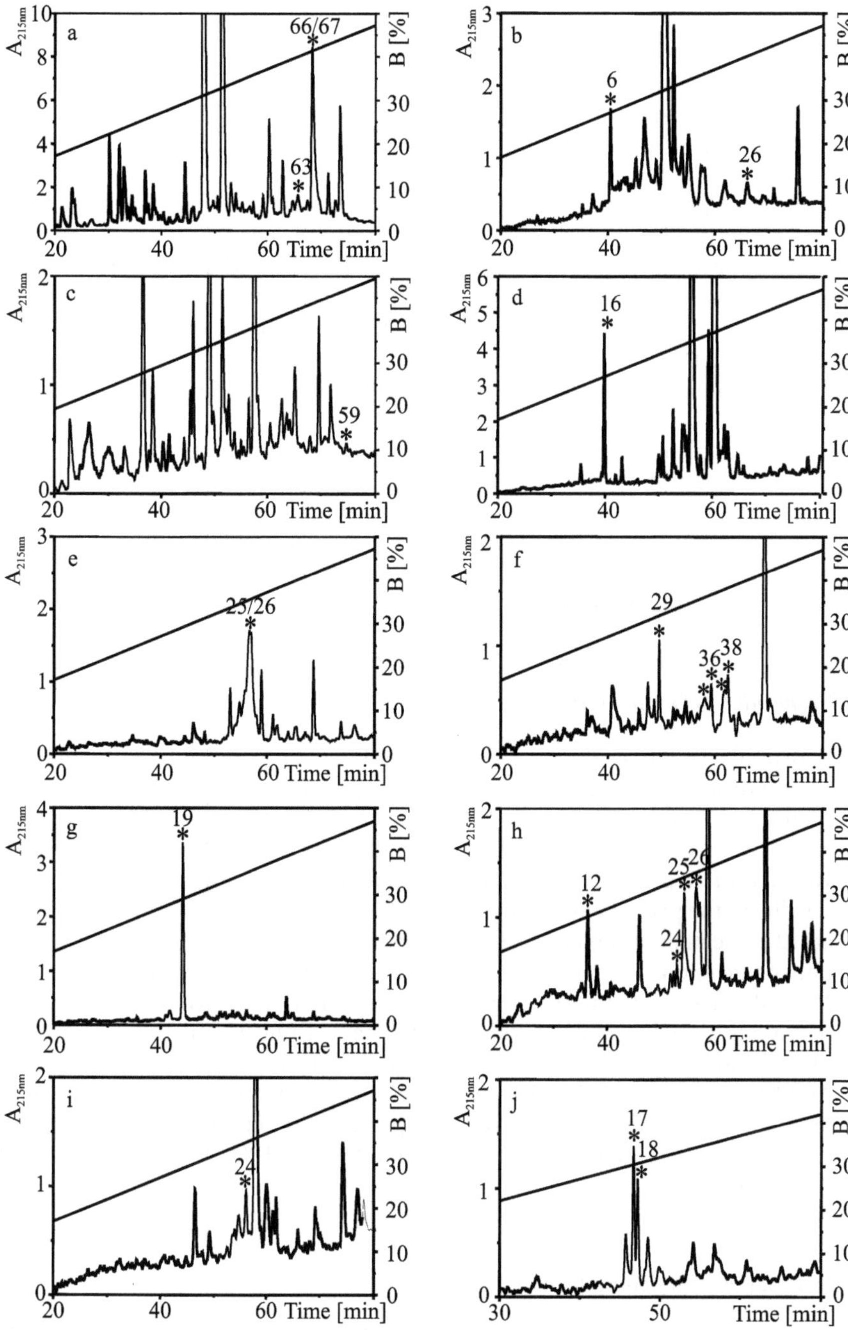

Fig. 17 a–j. Rechromatography of the fractions a–j (see Fig. 1). Fractions containing radioactivity are marked with an asterisk

In Fig. 18 the MALDI-PSD spectra of three phosphorylated peptides from the fractions a66, b6 and d16 are shown. The b- and y-ion series allow one to determine the structure of each peptide and to localize the phosphotyrosine residue in the primary structure of the peptides.

Phosphoserine and phosphothreonine can be identified due to the loss of phosphate which is only observed in reflection mode but not in the linear mode. Sometimes peptides undergo such a strong fragmentation that they are only detectable in the linear mode of a mass spectrometer (see Fig. 19).

Phosphoserine or -threonine residues can be localized in the same way. However, fragment ion spectra from peptides containing these *O*-phosphates show in most cases very intensive signals at $[M+H]^+$-80 Da (loss of HPO_3^{2-}) and $[M+H]^+$-98 Da (loss of $H_2PO_4^-$) [47–49]. This behaviour makes it sometimes impossible to localize the phosphoamino acid by ESI-MS/MS or MALDI-PSD experiments.

An example for the successful identification of phosphoserine and phosphothreonine peptides after radioactive labelling and micro-HPLC separation is given in Fig. 20. Both peptides undergo a strong phosphate cleavage which can be observed in the reflectron mode of the MALDI-TOF instrument. The interpretation of the fragment ion data is more difficult because a possible loss of 80 Da or 98 Da of each fragment ion have to be considered.

4.3.2
Localisation of Phosphohistidine

Phosphohistidine residues are unstable under acidic conditions and therefore a hexafluoroacetone/NH_3 gradient system at pH 8.6 is well suited. The common acetonitrile/water/TFA system is also possible for analysis but a fast chromatography is recommended, because the half-life of phosphohistidine is about 10–30 min at pH 3. An example for the identification of a phosphohistidine residue in the primary structure of HPr from *Bacillus subtilis* is given in Figs. 21 and 22. The major problem dealing with phosphohistidine is the instability of the phosphate group under acid conditions and during ionisation. The usage of MALDI mass spectrometry is only possible with basic matrices.

An additional example for the localisation of phosphohistidine is given by Medzihradsky et al. [50] with stability studies of the synthetic phosphopeptide Ac-SFTNPLHpSAAW-NH_2. However, the phosphorylation site cannot be determined to the histidine residue by mass spectrometry but to the partial sequence LHS. The major signals are derived from a loss of HPO_3^- and H_3PO_4 similar to the *O*-phosphates P-Ser and P-Thr.

4.3.3
Localisation of Acylphosphates

A direct identification of acylphosphates is difficult. However, selective reduction of acylphosphate with $NaBH_4$/[^3H]$NaBH_4$ to the corresponding alcohol and labelling with tritium is possible. After digestion and separation of tritium-containing peptides a structure determination can be done by mass spectrometry.

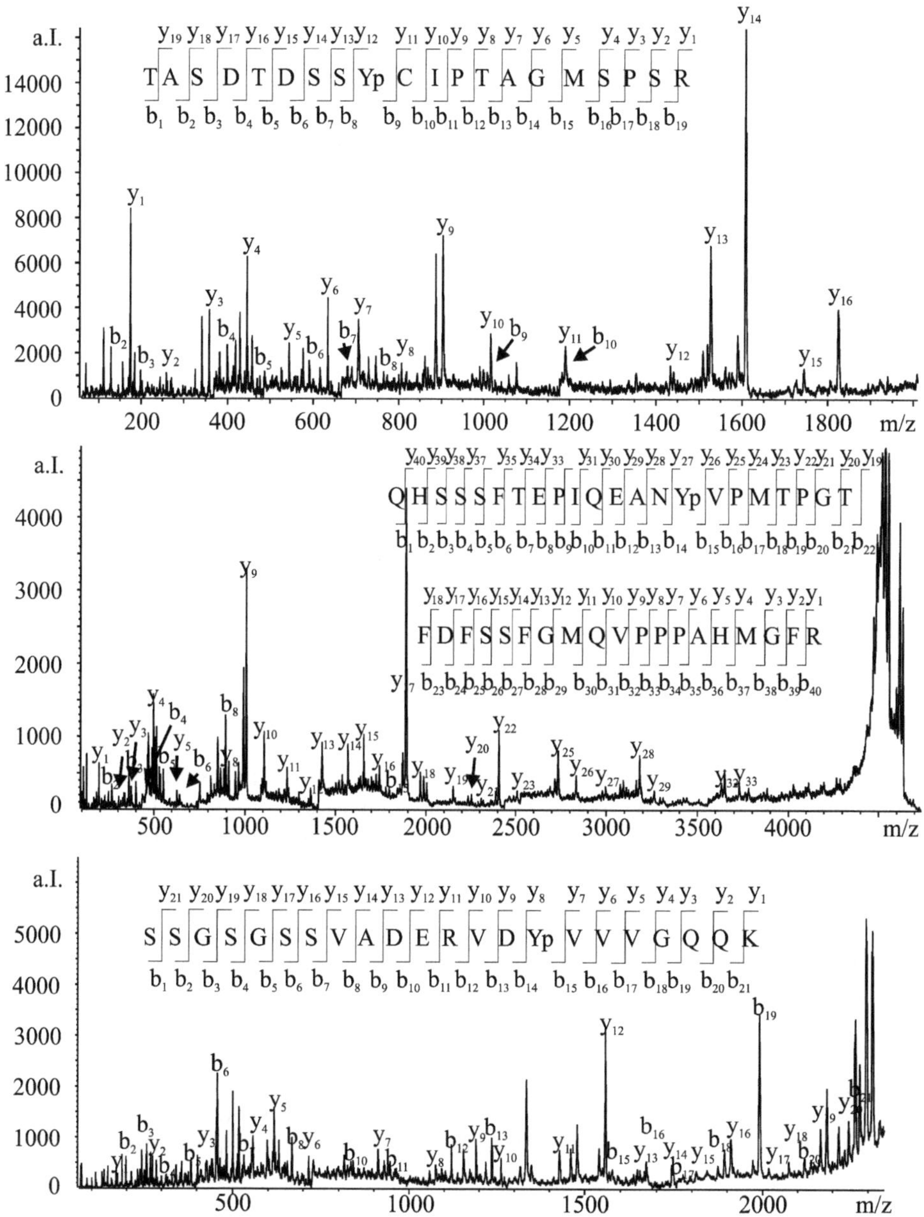

Fig. 18. MALDI-PSD spectra of tyrosine phosphorylated hGab-1. MALDI-PSD spectra of different phosphopeptides of the fractions a66, b6 and d16. *Top*: PSD spectrum of the peptide with *m/z* 2197.8 Da from fraction b6. The phosphorylated tyrosine residue (Y373) is easily located in the sequence TASDTDSSYpCIPTAGMSPSR. *Middle*: PSD spectrum of the peptide wit *m/z* 4367.0 Da from fraction a66. The phosphorylated tyrosine residue (Y472) is easily located in the sequence QHSSSFTEPIQEANYpVPMTPGTFDFSSFGMQ. *Bottom*: PSD spectrum of the peptide with *m/z* 2392.1 Da from fraction d16. The phosphorylated tyrosine residue (Y689) is easily located in the sequence SSGSGSSVADERVDYpVVVDQQK

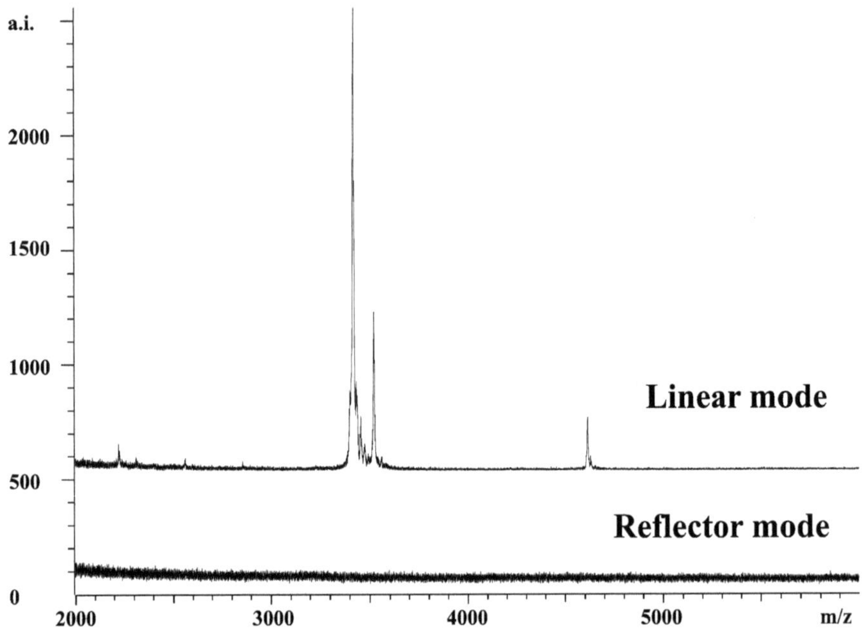

Fig. 19. Instable peptides are detectable in the linear mode of a MALDI-TOF. This peptide undergoes a complete fragmentation and is only detectable in the linear mode of the MALDI-TOF with a broad signal. However, such complete fragmentation is rather seldom

Sanders et al. [51] show the localisation of phosphoaspartate in CheY using $NaBH_4$ derivatisation followed by ESI tandem mass spectrometry. The reduction of phosphoaspartate leads to a homoserine residue which can easily be assigned in MS/MS spectra. An example for the identification of phosphoglutamate is given by Trumbore et al. [52].

4.3.4
Phosphocysteine

An example for the combination of Edman degradation and mass spectrometry is given by Weigt et al. [53] The phosphorylated EII^{Mtl} fragment from *Staphylococcus carnosus* is analysed with ESI-MS. After digestion with Glu-C followed by LC-MS analysis of the digest the fraction containing the phosphocysteine peptide was subjected to Edman degradation, allowing the positive identification of the phosphocysteine residue.

4.3.5
Acetylation

The two co-translational processes, cleavage of *N*-terminal methionine residue and N-terminal acetylation, are by far the most common modifications, occurring on the vast majority of eucaryotic proteins [54]. *N*-terminally acetylated pro-

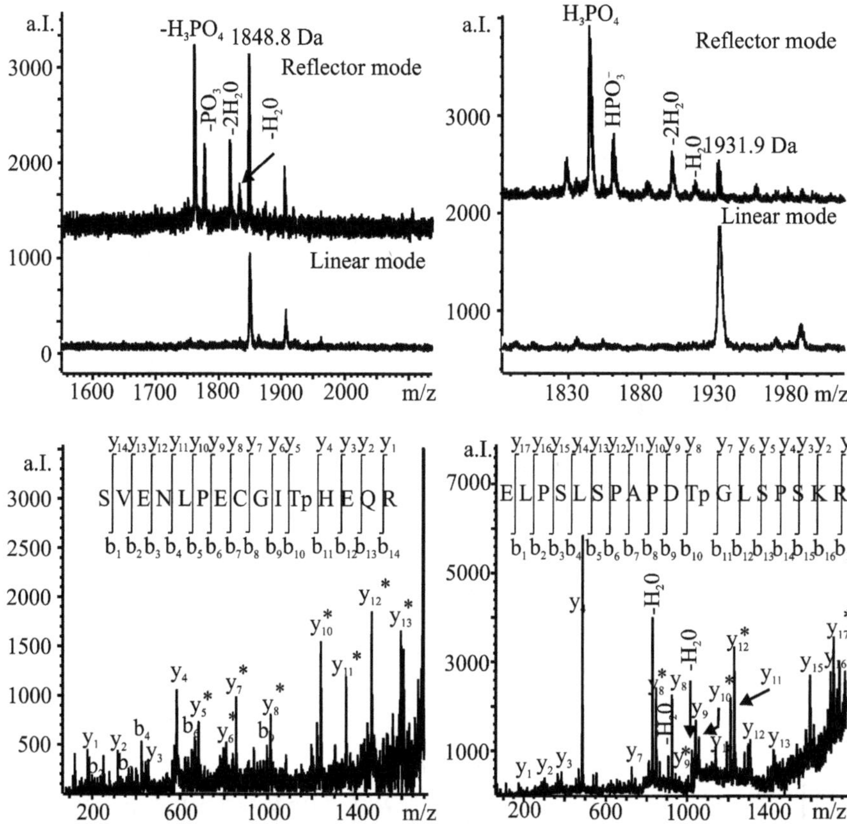

Fig. 20. Identification of two autophosphorylation sites of human PI4K92. *Top*: comparison of MALDI mass spectra recorded in the linear mode and reflectron mode. The cleavage of 80 Da and 98 Da is easily detected in the reflectron mode but not in the linear mode. *Bottom*: MALDI-PSD spectra of the isolated phosphopeptides. The phosphorylation site can be localized in both spectra. The peptide ELPSLSPAPDTpGLSPSKR contains more than only one residue which can be phosphorylated. However, the specific cleavage allow to localize the phosphate group to threonine 263

teins show a special behaviour in analysis of their tryptic peptides with mass spectrometry. The acetylated *N*-terminal peptide commonly shows a very high ionisation rate as demonstrated in Fig. 23.

The MALDI mass fingerprint unequivocally identifies mouse cyclophilin A. However, the base peak (2048.99 Da) in the mass spectrum cannot be explained by any unmodified peptide derived from the mouse cyclophilin A sequence. The ion is selected for a MALDI-PSD experiment and the generated fragmention spectrum identifies this peptide as the acetylated *N*-terminal peptide Ac-VNPTVFFDITADDEPLGR. The spectrum contains only y-ions due to the fact that the *N*-terminal NH_2-group cannot be protonated.

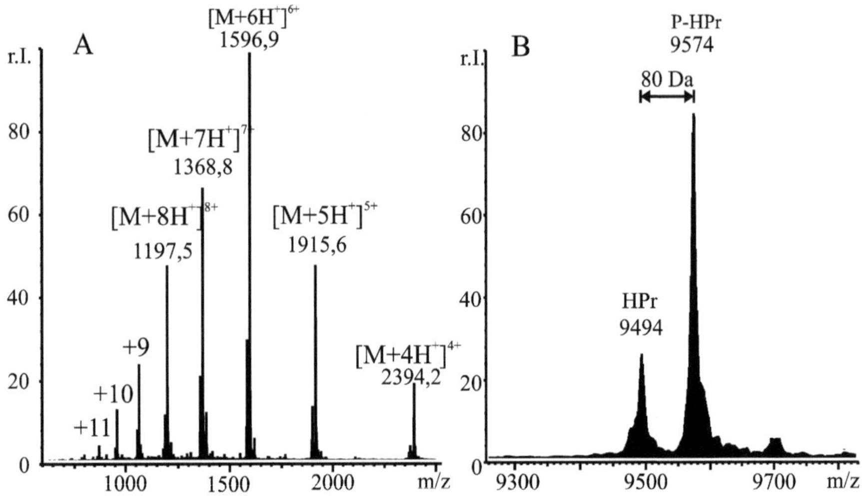

Fig. 21 A, B. Identification of phosphohistidine in the primary structure of HPr: **A** ESI nanospray spectrum of the histidine phosphorylated HPr before analysis. The charge states + 4 to + 11 can be observed; **B** deconvoluted ESI nanospray spectrum of the phosphorylated HPr. Unphosphorylated (9494 Da) and phosphorylated HPr (9574 Da) show a ratio of approximately 3 to 1. An aliquot of this sample was digested with trypsin at pH 7.8 for 4 h and analysed by LC-MS/MS (Fig. 22)

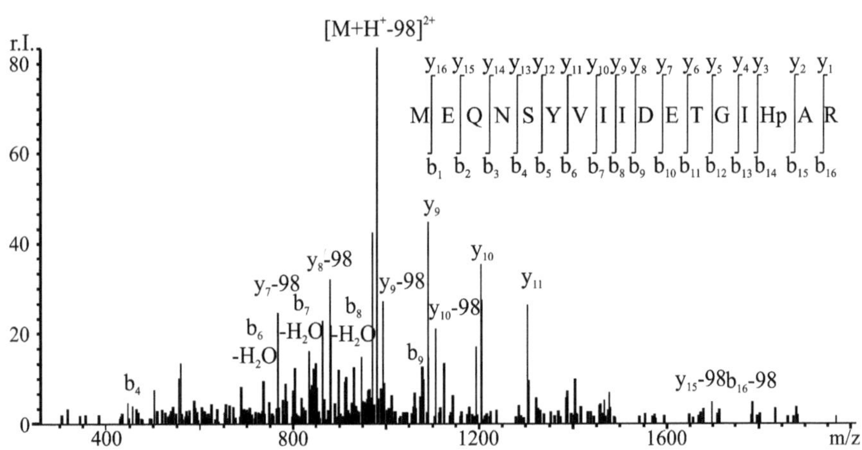

Fig. 22. ESI MS/MS spectrum of a phosphohistidine-containing peptide. ESI MS/MS spectrum of a histidine phosphorylated peptide from HPr. The phosphorylated amino acid can be localized with the b- and y-ions in the peptide MEQNSYVIIDETGIHpAR

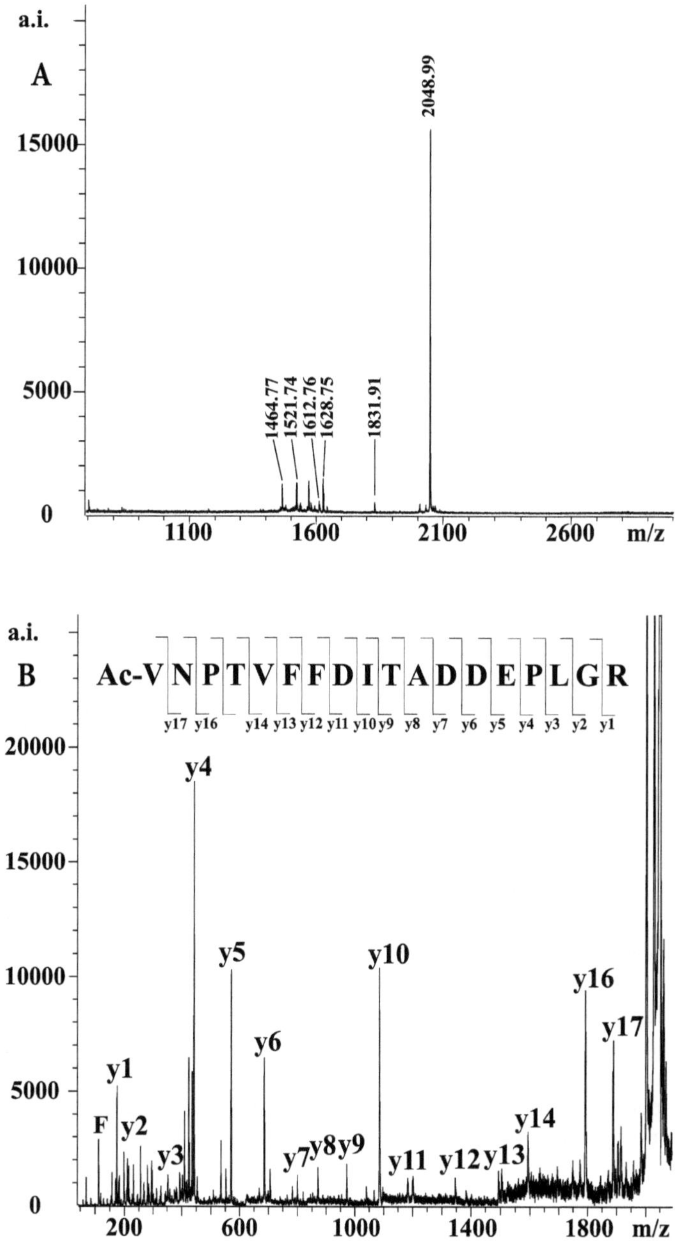

Fig. 23 A, B. Identification of an acetylated peptide: **A** MALDI mass fingerprint of a 2D gel spot. The protein cyclophilin A is identified by the peptide signals. However, the signal with *m/z* 2048.99 Da is not explained by any unmodified peptide of the protein; **B** MALD-PSD spectrum of the peptide ion with *m/z* 2048.99 Da. Only y-ions are obtained due to the fact that the acetylated *N*-terminus cannot be protonated any more. The nearly complete y-ion series is shown in the spectrum. I-type ions for phenylalanine (F, 120 Da) and C-terminal arginine (y_1, 175 Da) are detected

4.3.6
Modification Depending on Sample Handling – Oxidation and Alkylation

Methionine is easily oxidized during sample preparation prior to 2D-PAGE. Using the IPG method [55] a further modification is introduced in methionine-containing peptides. After isoelectric focusing and reduction/alkylation of proteins in the IPG gel methionine occurs at a 48 Da lower mass than before. Studies with the synthetic peptide YGGFMTSEK showed a loss of 48 Da located at methionine after incubation with IAA [56]. An oxidized methionine residue shows a further signal with lower resolution which results from a PSD fragment of the methionine sulfon side chain with approximately 59 Da lower mass to the oxidized peptide [57].

4.4
Protein Quantification Using Mass Spectrometry

Commonly, mass spectrometry is able to answer the question of which protein or proteins are in a sample, but unfortunately not of how much of a protein is in the sample. Several new approaches for quantitative mass spectrometry are described in the literature. A very promising approach – the isotope coded affinity tag (ICAT) method [58] – uses a cysteine modification with an isotope labelled biotin tag. For the relative quantification of two proteomes A and B the sample A is labelled with the "normal" ICAT reagents and sample B with the deuterium (D_8) ICAT reagents. After the modification both samples are combined and separated by 2D-PAGE or 2D-chromatography. Due to the fact that the ionisation probability of isotope coded peptides should be the same a relative quantification of the peptide amounts in sample A and B is possible. Using MALDI mass spectrometry doublet signals with a mass difference of 8 Da are generated. Two further methods use another approach for the isotope coding. An easy method is the cultivation of a microorganism onto a isotope medium like ^{15}N medium [59] and a control onto a "normal" medium. The cells are combined for further analysis and processed together. The other method is the incorporation of [^{18}O] [10]. Two [^{18}O] atoms are incorporated universally into the carboxyl termini of all tryptic peptides during a proteolytic digest of all proteins in the first pool. The second pool is treated in the some way only that [^{16}O] atoms are used (i. e. no labelling). The pooled and combined peptide mixtures are then separated and analysed by mass spectrometry, for a mass difference of 4 Da, for corresponding peptide pairs, and their isotope ratios.

5
References

1. Fenn JB, Mann M, Meng CK, Wong SF, Whitehouse CM (1989) Science 246:64
2. Karas M, Hillenkamp F (1988) Anal Chem 60:2299
3. Wilm M, Mann M (1996) Anal Chem 68:1
4. Chervet JP, Ursem D, Salzmann JP (1996) Anal Chem 68:1507
5. Schuerenberg M, Luebbert C, Eickhoff H, Lehrach H, Nordhoff E (2000) Anal Chem 72:3436

6. Spengler B, Kirsch D, Kaufmann R, Jaeger E (1992) Rapid Comm Mass Spectrom 6:105
7. Spengler B, Kirsch D, Kaufmann R (1992) J Phys Chem 96:9678
8. Shevchenko A, Wilm M, Mann M (1997) J Protein Chem 96:9678
9. Biemann K (1992) Annu Rev Biochem 61:977
10. Schnoelzer M, Jedrzejewski P, Lehmann W (1996) Electrophoresis 17:945
11. Shevchenko A, Chernushevich I, Ens W, Standing K, Thomson B, Wilm M, Mann M (1997) Rapid Comm Mass Spectrom 11:1015
12. Chait BT, Wang R, Beavis RC, Kent SB (1993) Science 262:89
13. Bartlet-Jones M, Jeffery WA, Hansen HF, Pappin DJ (1994) Rapid Comm Mass Spectrom 8:737
14. Sharon N, Lis H (1997) Glycoproteins: structure and function. In: H. J. Gabius and Gabius S (eds) Glycoscience: status and perspectives. Chapman & Hall, Weinheim, p 133
15. Schäffer C, Graninger M, Messner P (2001) Proteomics 1:248
16. Packer NH, Lawson MA, Jardine DR, Sanchez JC, Gooley AA (1998) Electrophoresis 19(6):981
17. Packer NH, Harrison MJ (1998) Electrophoresis 19(11):1872
18. Kuester B, Krogh TN, Mortz E, Harvey D (2001) Proteomics 1:350
19. Harvey DJ (1999) Mass Spectrom Rev 18:349
20. Harvey D (2001) Proteomics 1:311
21. Mock KK, Sutton CW, Cottrell JS (1992) Rapid Commun Mass Spectrom 6(4):233
22. Tsarbopoulos A, Bahr U, Pramanik BN, Karas M (1997) Int J Mass Spectrom Ion Process 169/170:251
23. Juhasz P, Costello CE, Biemann K (1993) J Am Soc Mass Spectrom 4:399
24. Kim YJ, Freas A, Fenselau C (2001) Anal Chem 73:1544
25. Allmaier G, Schäffer C, Messner P, Rapp U, Mayer-Posner FJ (1995) J Bacteriol 177:1402
26. Yamashita K, Ohkura T, Ideo H, Ohno K, Kanai M (1993) J Biochem (Tokyo) 114:766
27. Karas M, Bahr U, Dulcks T (2000) Fresenius J Anal Chem 366(6/7):669
28. Krogh TN, Berg T, Hojrup P (1999) Anal Biochem 274(2):153
29. Mortz E, Sareneva T, Haebel S, Julkunen I, Roepstorff P (1996) Electrophoresis 17(5):925
30. Cooper CA, Gasteiger E, Packer NH (2001) Proteomics 1:340
31. Sutton CW, O'Neill JA, Cottrell JS (1994) Anal Biochem 218(1):34
32. Kuster B, Naven TJ, Harvey DJ (1996) J Mass Spectrom 31(10):1131
33. Hanisch FG, Jovanovic M, Peter-Katalinic J (2001) Anal Biochem 290(1):47
34. Mirgorodskaya E, Hassan H, Clausen H, Roepstorff P (2001) Anal Chem 73(6):1263
35. Cooper CA, Packer NH, Redmond JH (1994) Glycoconjugate J 11:163
36. Naven TJ, Harvey DJ (1996) Rapid Commun Mass Spectrom 10(11):1361
37. Naven TJP, Harvey DJ (1996) Rapid Commun Mass Spectrom 10:829
38. Zhao Y, Kent SBH, Chait BT (1997) Proc Natl Acad Sci 94:1629
39. Gorisch H (1988) Anal Biochem 173(2):393
40. Mohr MD, Bornsen KO, Widmer HM (1995) Rapid Commun Mass Spectrom 9(9):809
41. Packer NH, Lawson MA, Jardine DR, Redmond JW (1998) Glycoconjugate J 15(8):737
42. Harvey DJ (1999) Mass Spectrom Rev 18:349
43. Domon B, Costello CE (1988) Glycoconjugate J 5:397
44. Meyer HE, Hoffmann-Posorske E, Korte H, Covey T, Donella-Deana A (1991) In: Heilmeyer LMG (ed) Cellular regulation by protein phosphorylation. NATO ASI Series, vol H56:43
45. Meyer HE, Eisermann B, Donella-Deana A, Perich JW, Hoffmann-Posorske E, Korte H (1993) Protein Sequences Data Anal 5:197
46. Meyer HE, Hoffmann-Posorske E, Donella-Deana A, Korte H (1991) In: Hunter T, Sefton BM (eds) Methods in enzymology protein phosphorylation, vol 201, part B. Academic Press, New York, p 206
47. Butt E, Bernhardt M, Smolenski A, Kotsonis P, Frohlich LG, Sickmann A, Meyer HE, Lohmann SM, Schmidt HH (2000) J Biol Chem 275(7):5179
48. Sickmann A, Marcus K, Schafer H, Butt-Dorje E, Lehr S, Herkner A, Suer S, Bahr I, Meyer HE (2001) Electrophoresis 22(9):1669

49. Sickmann A, Meyer HE (2001) Proteomics 1:200
50. Medzihradszky KF, Phillips NJ, Senderrowicz L, Wang P, Turck CW (1997) Protein Sci 6:1405
51. Sanders DA, Gillece-Castro BL, Stock AM, Burlingame AL, Koshland DE (1989) J Biol Chem 264:21,770
52. Trumbore M, Wang RH, Enkemann SA, Berger SL (1997) J Biol Chem 272:26,394
53. Weigt C, Korte H, Pogge von Strandmann R, Hengstenberg W, Meyer HE (1995) J Chromatogr A 712:141
54. Polevoda B, Sherman F (2000) J Biol Chem 275(47):36,479
55. Bjellqvist B, Righetti PG, Gianazza E, Görg A, Westermeier R, Postel W (1982) J Biochem Biophys Methods 6:317
56. Sickmann A, Dormeyer W, Wortelkamp S, Woitalla D, Kuhn W, Meyer HE (2000) Electrophoresis 21:2721
57. Schnölzer M, Lehmann WD (1997) Mass Spectrom Ion Process 169/170:263
58. Gygi SP, Rist B, Gerber SA, Turecek F, Gelb MH, Aebersold R (1999) Nat Biotechnol 17(10):994
59. Oda Y, Nagasu T, Chait BT (2001) Nat Biotechnol 19(4):379

Received: April 2002

Adv Biochem Engin/Biotechnol (2003) 83: 177–187
DOI 10.1007/b11116

Protein Arrays and Their Role in Proteomics

Dolores J. Cahill[1] · Eckhard Nordhoff[2]

[1] Max-Planck-Institute for Molecular Genetics, Ihnestrasse 73, 14195 Berlin, Germany
E-mail: cahill@molgen.mpg.de
[2] Scienion AG, Volmerstrasse 76, 12489 Berlin, Germany
E-mail: nordhoff@scienion.de

Arraying technologies have shown the way to smaller sample volumes, more efficient analyses and higher throughput. Proteomics is a field, which has grown in significance in the last five years. This review outlines recent developments in protein arrays and their applications in proteomics, and discusses the requirements, current limitations and the potential and future perspectives of the technology.

Keywords. Microarray, Protein chips, Proteomics, Mass spectrometry, cDNA expression library

1 Introduction

Proteomics can be most broadly defined as the systematic analysis of all proteins contained in a biological sample, which aims to document their identity, expression level and localization in space and time. Ultimately, the goal is to elucidate their relationships and functional networks [1, 2]. High-density DNA, protein and antibody chips provide a miniaturized platform for molecular biological and molecular medical research and enable experiments hitherto impossible in scale, time, and detection sensitivity. This review outlines current techniques used in the generation and analysis of protein arrays, and their applications in proteomics.

2
Protein Arrays – Overview

Since proteins translate genomic sequence information into function, thereby enabling biological processes, a full understanding of the expression profile of a tissue or organism on both the genomic and proteomic levels ideally requires the screening of many samples in parallel, as rapidly as possible. Here we review protein arrays, and how the analysis of recombinant proteins used to generate proteins arrays may be analysed by mass spectrometry. The data generated may be used not only to assess the purity of these proteins and to identify them, where necessary, but we also discuss how the areas of protein arrays and proteomics may increasingly complement each other.

Protein (and DNA) arrays are ordered arrangements of individual samples, enabling their parallel analysis, and which are used for biological experimentation. The samples to be arrayed are generally in microtitre plates. The 96-well microtitre plate is still the most widely used in immunoassays, whereas 384-well plates are taking over for many assays and are now standard for storage and handling of clone libraries. There are 1536-well plates now available, and etched glass or silicon wafers make substrates for "nanoplates" (e.g. 9600-wells) [3]. Until recently it had not been possible to analyse proteins using the same high-density array, automated approach described for the DNA arrays. We have applied automation technologies to the high-throughput, large-scale analysis of proteins, by generating cDNA expression libraries, high-density protein arrays [4, 5] and micro-arrays [6]. A protein array can be generated using two methods, either by arraying and growing clones and subsequent induction of protein expression on a membrane or filter [4] or by expressing each proteins from the protein expressing set in microtitre plates, and arraying the purified proteins onto filters [6], or more recently on glass slides (see below). Further developments in protein arrays include the generation of low-density protein arrays on filter membranes, such as the universal protein array system (UPA), which is based on the 96-well microtitre plate format [7]. Protein micro-arrays have also been printed, again in the microtitre well format, on an optically flat glass plate containing 96 wells formed by an enclosing hydrophobic Teflon mask [8]. Inside the wells, arrays of 144 (4×36) elements each were spotted using a 36-capillary-based print head attached to a precise, high-speed, X-Y-Z robot. Standard ELISA techniques and a scanning CCD detector were employed for imaging of arrayed antigens. An array immunosensor has been developed which allowed the simultaneous detection of clinical analytes [9]. This is also in a microtitre plate format and had capture antibodies and analytes arrayed onto microscopic slides using flow chambers in a cross-wise fashion. Detection was again via fluorescent labels and CCD-based optical readout. Although still at a low-density stage (6×6 pattern), the technique has high-throughput potential as it involves automated image analysis and micro-fluidics which is already becoming one of the future formats for enzyme activity and other assays [10] (Caliper Technologies, Mountain View, CA, USA, www.calipertech.com; Orchid Biocomputer Inc, Princeton, NJ, USA, www.orchid.com).

To manufacture three-dimensional arrays on a flat surface, a gel photo- or persulfate-induced co-polymerisation technique has been developed to produce oligonucleotide, DNA and protein microchips on polyacrylamide gel pads from 10×10 to 100×100 μm, separated by a hydrophobic glass surface [11]. The three-dimensional polyacrylamide gel provides a more than 100 times greater capacity for immobilisation than does a two-dimensional glass support, thus increasing the sensitivity of measurements considerably [12]. A method for fabricating antibody arrays has been reported which uses a micro-moulded hydrogel "stamper" and an aminosilylated receiving surface [13]. The stamper deposits protein as a submonolayer, as shown by I-125 labelling and atomic force microscopy, whilst antibody activity was retained. Other approaches to protein microarrays have been reported using either photolithography of silane monolayers [14] or gold [15], combining microwells with microsphere sensors [16] or inkjetting onto polystyrene film [17]. Those advances are focused on the fabrication of miniaturised immunoassay formats by patterning of single proteins (e.g. BSA, avidin or monoclonal antibodies). Joos et al. have described the application of a microarray-based immunoassay, which allows the simultaneous quantification of autoantibodies in the sera of autoimmune patients, using 18 known autoantigens [18]. De Wildt et al. have described a method to generate high content antibody arrays using recombinant single-chain variable fragment (scFv) antibodies [19].

MacBeath and Schreiber have used a high-precision contact-printing robot to generate protein arrays on aldehyde and BSA-*N*-hydroxysuccinimide treated glass slides, and then performed proof-of-principle experiments with protein-protein interactions, and identifying substrates of protein kinases and protein targets of small molecules (DIG, biotin and streptavidin) on these slides [20].

Uetz et al. reported the most complete yeast two-hybrid screen for the identification of protein-protein interactions, where approximately 5300 yeast proteins were individually fused to transcriptional activation domains which were systematically probed with yeast proteins fused to a DNA-binding domain. This screen was performed in an array format resulting in the identification of 957 potential protein-protein interactions [21]. Zhu et al. cloned 5800 yeast open reading frames into a yeast expression vector, where the expressed proteins were expressed as fusion proteins, with a glutathione *S*-transferase-polyhistidine tag at their NH2-terminal. The proteins were purified and arrayed on nickel coated microscope slides. These slides were used to study protein-protein interactions, where new calmodulin- and phospholipid-interacting proteins were identified, as well as post-translational modifications and protein-drug interactions [22].

For capturing and analysis of specifically labelled proteins, ligand-coated surfaces are available, like the SELDI ProteinChip (Ciphergen Biosystems Inc., Palo Alto, CA, USA, www.ciphergen.com), XNA on Gold (INTERACTIVA Biotechologie GmbH, Ulm, Germany, www.interactiva.de) and various BIAcore chips (Biacore AB, Uppsala, Sweden). Currently, however, the SELDI and the BIAcore chips provide only four or eight sample positions for parallel experimentation, and therefore are not further discussed here.

Tissue micro-arrays have been developed for high-throughput molecular profiling of tumour specimens [23], which involved a robot punching cylinders (0.6 mm wide, 3–4 mm high) from 1000 individual tumour biopsies embedded in paraffin and arraying them in a 45×20 mm paraffin block (commercially available from Beecher Instruments, Silver Spring, MD, USA, www.beecherinstruments.com). On serial sections, tumours are then analysed in parallel by immunohistochemistry, fluorescence in situ hybridisation (FISH) and RNA/RNA in situ hybridisation. This system allowed the microscopic scanning of an immunohistochemistry array slide containing 645 specimens in less than two hours. It should be of great help for the simultaneous analysis of tumours from many different patients at different stages of disease to establish the therapeutic importance of new candidate marker genes more rapidly [24].

3 Generation of Protein Expression Libraries as a Resource for Protein Arrays

To generate protein arrays with large numbers of proteins, the main requirement is the source of these proteins. Therefore, the main requirement is to provide the protein content to put on the arrays, which we do by generating cDNA libraries (reviewed [25]) in a protein expression vector. We have generated a non-redundant, human Unigene-Uniprotein set from a human brain cDNA library [4]. This set of clones contains over 10,000 clones that express different (non-redundant) human proteins derived from a human brain cDNA expression library. These proteins have been arrayed on membranes and on glass slides (protein chips). We have screened these high density, high content protein arrays with antibodies and serum from patients with autoimmune disease. Currently we are testing different surfaces (glass, plastic, coated surfaces) to optimise the protein chip microscope slide format. Also, antibody arrays are being tested on these chip surfaces, with the aim to develop chip-based assays and for possible applications in diagnostics, and perhaps in the future for detection of proteins in lysates, or in a sandwich assay format. The major current drawback of antibody arrays is the source of stable binders or antibodies, where the binding specificity (or cross-reactivity) of the antibody is known.

The second required development was to express and purify thousands of proteins in parallel, which will be used to generate the protein arrays (reviewed [26]). We have moved from the initial *E. coli* expression system [4] to developing expression vectors and systems, which are appropriate for high throughput in *Pichia pastoris* [27] and in *S. cerevisae* [28]. We are currently also generating a system to get improved expression of secreted proteins in *E. coli.*

4 Characterization of Protein Expression Libraries – Quality Control

We have shown that cDNA libraries can be screened for protein expression on high-density filter membranes, which described the use of high-density filters for parallel DNA hybridisation, protein expression and antibody screening [4]. Such approaches make a large number of different protein products encoded by cDNA clones available for test experiments. We have also seen that when such protein arrays are screened with serum from patients with auto-immune diseases, a subset of proteins can be identified on the array which can correlate with the disease. Such proteins require extensive characterisation, and large numbers of patients would be required to be screened. However, these selected proteins may be candidate diagnostic markers for the auto-immune disease screened. These experiments can generate a wealth of valuable information in a short time based on which promising subsets can be selected and used for the generation of application-specific protein arrays, or further characterized as described below.

Generation of high-quality high-density protein arrays requires an efficient approach to protein purification, identification and characterization of their quality [29], as well as rapid arraying of these proteins. A MALDI time-of-flight (TOF) mass spectrometry based technique has been described for the quality control of proteins expressed in 96-well plates [29–31]. In these experiments clones were selected from the hEx1 human foetal brain cDNA expression library and all necessary sample handling from *E. coli* cell culture to the mass spectrometric analyses were performed in 96-well microtitre plates. The library was constructed in an expression vector that allows expression of *N*-terminal MRGS$(His)_6$-tag fusion proteins. The expressed proteins were metal-chelate affinity purified under denaturing conditions using Ni-NTA agarose as the affinity matrix. The bound MRGS$(His)_6$-tag fusion proteins were purified using two different washing buffers, the first one optimised for removal of non-specifically bound proteins as well as other biomolecules present in the host cells, and a second, low-salt buffer, optimised for MALDI-TOF-MS. To identify and characterise the quality of the purified proteins, these were analysed in two different ways.

In one experiment the bound proteins were digested in situ with the protease trypsin, and the released cleavage peptides were recovered and mass analysed by MALDI-TOF-MS. The recorded peptide molecular mass fingerprints were then used to identify these proteins. Three examples are shown in Fig. 1 [30]. This approach, called MALDI-TOF-MS peptide mapping, has become an important analytical tool for the rapid identification of small amounts of native proteins [32–37]. Typically, the proteins of interest are first separated by two-dimensional gel electrophoresis or affinity enrichment followed by SDS-PAGE, the stained spots or bands are excised, the enclosed proteins are digested in situ with a specific protease, most frequently trypsin, and the resulting peptides are mass analysed. These data are compared with expected values computed from sequence database entries according to the enzyme's cleavage specificity. Finally, the results are scored and the ranking suggests the protein being identified or not.

Mass spectrometric peptide mapping, combined with the high detection sensitivity of state-of-the-art MALDI-TOF mass spectrometers and their capability to analyse fully automatically many samples in a short time, has made possible large-scale proteome research projects. Compared to native proteins, which can be the subject of many different post-translational modification reactions that alter their primary structure and thereby render the identification difficult or

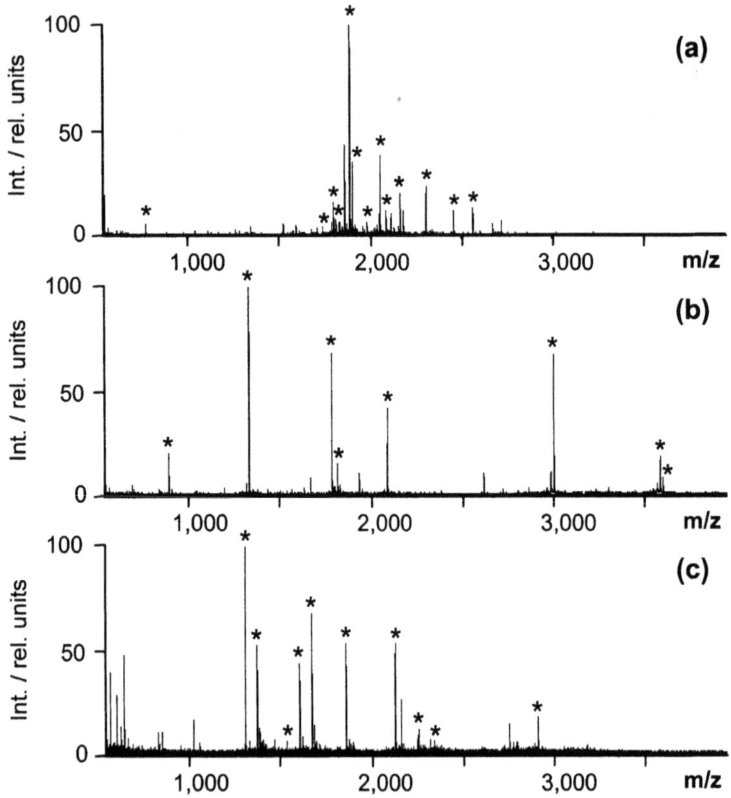

Fig. 1 a–c. MALDI-TOF-MS tryptic peptide mass maps of MRGS(His)6-tag fusion proteins of the hEx1 cDNA expression library. Following nickel-chelate affinity purification, the bound proteins were digested in situ with the protease trypsin and the cleavage peptides were mass analyzed. Based on these data, the purified proteins were unambiguously identified in the NCBI sequence database (release, September 2001): **a** PDZ domain protein 3′ variant 4; **b** proteasome subunit, alpha type 1 and **c** human bleomycin hydrolase. * signals that were assigned to tryptic cleavage products of the identified protein

impossible, the identification of recombinant proteins expressed in bacterial host cells is simple and straightforward. In turn, the above strategy appears well suited to screen large cDNA expression libraries for interesting clones or to identify all clones that produce the encoded fusion protein. Compared to DNA sequencing or other DNA based methods, this approach has the advantage that the identification is performed on the protein level; hence, clones that do not express the recombinant gene are discarded. Clones that quickly degrade the expressed polypeptide chains are discarded too. These problems are difficult or impossible to recognize and sort out solely based on DNA code. The obvious strategy is first to screen on the protein level and then determine the DNA sequence of positives clones for a detailed characterization. In a second experiment, the affinity purified MRGS$(His)_6$-tag fusion proteins were not digested with trypsin. Instead, they were recovered from the stationary phase and a small aliquot (0.5 µl) of the eluate was analysed by MALDI-TOF-MS [29]. Figure 2 shows seven examples of the mass spectrometric data obtained from different clones of the hEx1 expression library. A second, larger aliquot (5 µl) was analysed by SDS-PAGE. From the obtained mass spectra the homogeneity and exact molecular weight was determined for each protein. The gel electrophoretic data were used to confirm their purity. All these data can be generated in a short time for many samples and provide a solid basis to judge the quality of the purified products.

In the above protocol, the use of MALDI-TOF-MS and SDS-PAGE is complementary. The first technology yields accurate molecular masses (accuracy better than 100 ppm), which can be used to determine the exact length of the expressed sequence. In addition, MALDI-TOF-MS can resolve protein species that differ in their primary structure only by a few atoms, and thereby reveal secondary modifications, e.g. oxidation of methionine or trytophane residues, incomplete reduction of cysteine-glutathione disulphide bridges (Fig. 2), or incomplete removal of *N*-terminal methionine. Furthermore, MALDI-TOF-MS can detect contaminants that are smaller than the cloned product, especially peptides (<10 kDa), at a sensitivity level (low femtomol/picogram range) unsurpassed by any other analytical technique. However, the detection sensitivity of MALDI-TOF-MS strongly declines with increasing molecular weight and abundant products of lower molecular weight negatively affect the detection of higher molecular weight components (signal suppression). This hampers or, in the worst case, excludes the detection of contaminants or by-products whose molecular masses exceed the molecular mass of the correct product. That limitation is compensated for by the additional use of SDS-PAGE, because in the subsequent Coomassie blue or silver staining procedure, with respect to the number of molecules, the detection sensitivity for proteins strongly increases with increasing molecular weight, and in general smaller proteins do not affect the detection of larger proteins.

Possible contaminants include peptides, proteins or other molecules of the host cell background, degradation products of the expressed fusion proteins, especially those fragments that harbour the affinity tag, as well as expression products whose primary structure has been modified during their expression, lysis of the cells or their purification. These can be detected as described above and in many cases identified solely based on their molecular mass determined by mass spectrometry or, in case of satellite signals, based on determined mass differences that are compared to expected mass shifts caused by known modifications. Examples for this are shown in Fig. 2. In this case, the SDS-PAGE analysis revealed additional bands whose identity is unclear; these can be excised and the contained proteins identified by MALDI-TOF-MS peptide mapping. Once identified, the next step is to calculate the molecular mass of the sequence retrieved from the database as well as *N*- and *C*-terminally shortened and, if necessary, all other possible subsets of it, and to compare these data with the molecular masses determined by MALDI-TOF-MS. It follows that the combination of MALDI-TOF-MS peptide mapping and the analysis of the intact proteins by mass spectrometry and by SDS-PAGE supplement each other. Based on the generated information and the required chip design and its applications, the scientist can decide for each protein whether a single-pass metal-chelate affinity purification

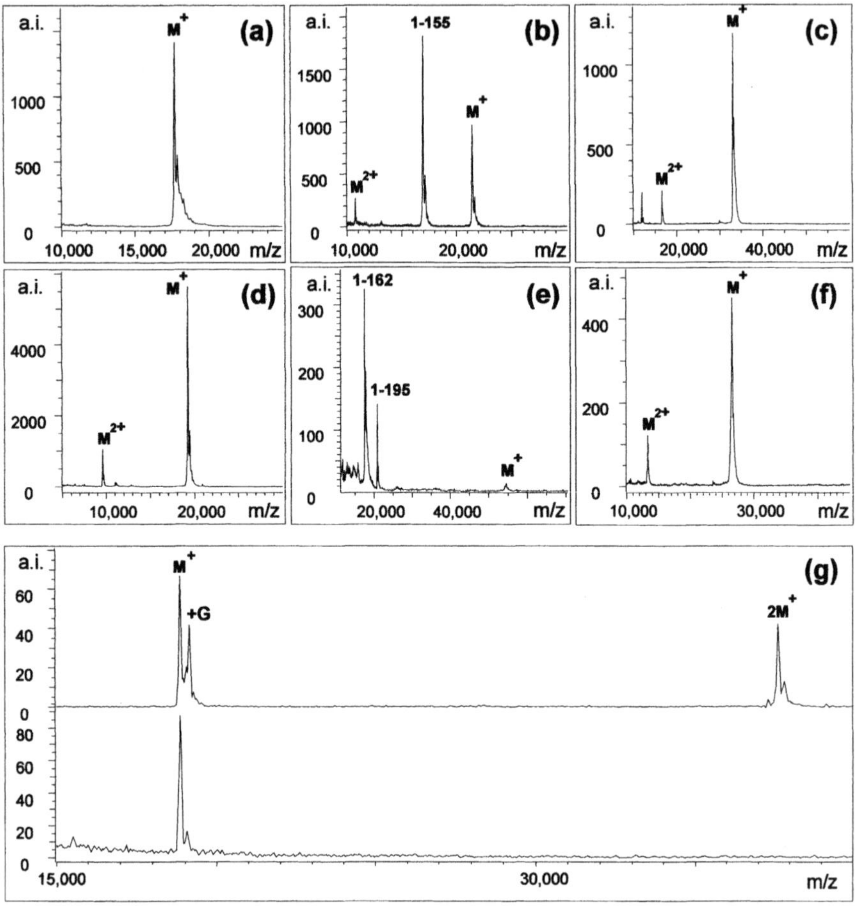

Fig. 2a–g. MALDI-TOF-MS of MRGS(His)$_6$-tag fusion proteins of the hEx1 cDNA expression library. Following nickel-chelate affinity purification, the obtained eluates were mass analysed; M^+ and $M2^+$, singly and doubly charged molecular ions of expected expression products. The numbers of the first and last amino acid indicate assigned *C*-terminal truncated protein sequences: **a** 40S ribosomal protein S25; **b** peptidyl-prolyl *cis-trans* isomerase A; **c** 40S ribosomal protein S3A; **d** tubulin alpha-1 chain; **e** beta tubulin; **f** golgin-95. For some expression products the determined molecular masses exceeded the predicted values by 300 Da indicative of glutathionylation, and for one clone, expressing cDNA of CMP-*N*-acetylneuraminante-beta-1,4-galactoside alpha-2,3-sialyltransferase, in addition a strong protein-dimer molecular ion signal was recorded indicative of protein-protein disulphide bridges. These indications were verified by reduction with DTT prior to the mass spectrometric analysis; **g** mass spectra obtained from this clone before (*upper spectrum*) and after (*lower spectrum*) reduction with DTT. +G, single glutathionylation, $2M^+$, singly charged protein dimmer

is sufficient, this purification is not suited or an additional purification step is needed (e.g. size exclusion chromatography for the removal of proteolytic degradation products).

5
Applications of Protein Arrays in Proteomics

Protein arrays enable high-throughput screening for gene expression and molecular interactions. The main advantage of a set of recombinant proteins is that this approach achieves the large-scale systematic provision of recombinant proteins for functional studies by making and arraying cDNA expression libraries and by allowing the direct connection from DNA sequence information on individual clones to protein products and back again on a whole genome level. We have proposed an application of these recombinant proteins in proteomics [38]. We have recently shown that the spectra derived from proteins from 2D gels and recombinant proteins can be used to compare and identify proteins from these sources [39]. This approach makes translated gene products directly amenable to high-throughput experimentation and generates a direct link between protein expression and DNA sequence data. Since protein micro-arrays are a useful tool for connecting gene expression analysis and molecular binding studies on a whole-genome level, if differentially expressed genes are identified using cDNA micro-arrays, the same clones can be analysed simultaneously for protein expression in different cellular systems, by mass spectrometry or by in vitro transcription/translation. On identical protein micro-arrays, expression clones can be screened for binding to other proteins (e.g. antibodies) or to diverse molecules from nucleic acids to small-molecule ligands. This versatility makes protein micro-arrays a promising multi-purpose tool for diagnostic use. We have described their use for ligand-receptor interaction studies, diagnosis and antibody specificity characterisation [3]. For example, the low-density universal protein array system (UPA) has been used for studying interactions with protein, DNA, RNA and small chemical ligand probes [7]. High-density protein filters of the hEx1 library were screened with antibodies against the human proteins GAPDH and HSP90 alpha, and positives were confirmed by cDNA probing and sequencing [4]. hEx1 protein filters have now been used by a number of groups to detect expression clones using antibodies, single-chain Fv fragments or other protein-protein or nucleic acid-protein interaction screening [3]. Lueking et al. have shown that protein microarrays enable sensitive antibody specificity screening [27]. The hEx1 library is currently being used for profiling of sera from patients with alopecia and rheumatoid arthritis, along with sex-matched, age-matched normal sera. It is envisaged that such screenings will generate novel diagnostic targets, which have applications as a method of characterising the particular disease based on the antibody profile of the patient, where such protein arrays could then be used as a diagnostic or prognostic tools.

The interest and application of a high-throughput approach is focussed in finding all genes, their in vivo functions and the features of the corresponding proteins. Information on a gene's expression is important for its potential exploitation. A gene's expression can be highly specific to a tissue, organ, cell type or disease and, as such, may be attractive as targets for development of highly specific therapeutics and diagnostics. Even a gene of unknown function may have medical utility if its expression pattern can be determined. Diagnostic tests can be developed for a disease marker if its presence or absence can be detected. To achieve this goal, methods and technologies operating reliably with many samples in high-throughput and in parallel are major requirements.

Such approaches make protein products encoded by cDNA clones available for the generation of specific, well-defined protein arrays. This requires a highly parallel approach to protein expression analysis, including the simultaneous expression of large numbers of cDNA clones in an appropriate vector system and high-speed arraying of protein product [38].

6
Future Prospects

Protein arrays appear as new and versatile tools in functional genomics and proteomics. An ever-increasing variety of array formats becomes available, e.g. microtitre plates, patterned arrays, three-dimensional pads, flat surface spot arrays or micro-fluidic chips. DNA and proteins can be arrayed onto different surfaces, e.g. membrane filters, polystyrene film, glass, silane or gold. Increasing the output from protein chips involve increasing the quality of the data output, high speed arraying, high throughput hybridisation devices with on-line detection, plus integrated image and data analysis tools.

7
References

1. Blackstock W, Weir MP (1999) Trends Biotechnol 17:121
2. Banks RE, Dunn MJ, Hochstrasser DF, Sanchez JC, Blackstock W, Pappin DJ, Selby PJ (2000) Lancet 356:1749
3. Walter G, Büssow K, Cahill D, Lueking A, Lehrach H (2000) Curr Opin Microbiol 3:298
4. Büssow K, Cahill D, Nietfeld W, Bancroft D, Scherzinger E, Lehrach H, Walter G (1998) Nucl Acids Res 26:5007
5. Cahill D, Büssow K, Walter G, Lehrach H (1998) Patent PCT Serial No. WO 09/070,590
6. Lueking A, Horn M, Eickhoff H, Büssow K, Lehrach H, Walter G (1999) Anal Biochem 270:103
7. Ge H (2000) Nucl Acids Res 28, e3 (I–VII)
8. Mendoza LG, McQuary P, Mongan A, Gangadharan R, Brignac S, Eggers M (1999) Biotechniques 27:778
9. Rowe CA, Scruggs SB, Feldstein MJ, Golden JP, Ligler FS (1999) Anal Chem 71:433
10. Cohen CB, Chin DE, Jeong S, Nikiforov TT (1999) Anal Biochem 273:89
11. Guschin D, Yershov G, Zaslavsky A, Gemmell A, Shick V, Proudnikov D, Arenkov P, Mirzabekov A (1997) Anal Biochem 250:203
12. Parinov S, Barsky V, Yershov G, Kirillov E, Timofeev E, Belgovskiy A, Mirzabekov A (1996) Nucl Acids Res 24:2998
13. Martin BD, Gaber BP, Patterson CH, Turner DC (1998) Langmuir 14:3971
14. Mooney JF, Hunt AJ, McIntosh JR, Liberko CA, Walba DM, Rogers CT (1996) Proc Natl Acad Sci USA 93:12,287
15. Jones VW, Kenseth JR, Porter MD, Mosher CL, Henderson E (1998) Anal Chem 70:1233
16. Michael KL, Taylor LC, Schultz SL, Walt DR (1998) Anal Chem 70:1242

17. Ekins RP (1998) Clin Chem 44:2015
18. Joos TO, Schrenk M, Hopfl P, Kroger K, Chowdhury U, Stoll D, Schorner D, Durr M, Herick K, Rupp S, Sohn K, Hammerle H (2000) Electrophoresis 21:2641
19. de Wildt RM, Mundy CR, Gorick BD, Tomlinson IM (2000) Nat Biotech 18:989
20. MacBeath G, Schreiber SL (2000) Science 289:1760-1763
21. Uetz P, Giot L, Cagney G, Mansfield TA, Judson RS, Knight JR, Lockshon D, Narayan V, Srinivasan M, Pochart P, Qureshi-Emili A, Li Y, Godwin B, Conover D, Kalbfleisch T, Vijayadamodar G, Yang M, Johnston M, Fields S, Rothberg JM (2000) Nature 403:623
22. Zhu HBM, Bangham R, Hall D, Casamayor A, Bertone P, Lan N, Jansen R, Bidlingmaier S, Houfek T, Mitchell T, Miller P, Dean RA, Gerstein M, Snyder M (2001) Science 293:2101
23. Kononen J, Bubendorf L, Kallioniemi A, Barlund M, Schraml P, Leighton S, Torhorst J, Mihatsch MJ, Sauter G, Kallioniemi OP (1998) Nat Med 4:844
24. Theillet C (1998) Nat Med 4:767
25. Clark MD, Panopoulou GD, Cahill DJ, Büssow K, Lehrach H (1999) Meth Enzymol 303:205
26. Cahill DJ (2001) J Immunol Meth 250:81
27. Lueking A, Holz C, Gotthold C, Lehrach H, Cahill DJ (2000) Protein Express Purif 20:372
28. Holz C, Lueking A, Bovekamp L, Gutjahr C, Bolotina N, Lehrach H, Cahill DJ (2001) Genome Res 11:1730
29. Büssow K, Nordhoff E, Lübbert C, Lehrach H, Walter G (2000) Genomics 65:1
30. Egelhofer V, Büssow K, Luebbert C, Lehrach H, Nordhoff E (2000) Anal Chem 72:2741
31. Gobom J, Schuerenberg M, Theiss D, Mueller M, Lehrach H, Nordhoff E (2001) Anal Chem 73:434
32. Henzel WJ, Billeci TM, Stultz JT, Wong SC, Grimley C, Watanabe C (1993) Proc Natl Acad Sci USA 90:5011
33. Mann M, Roepstorff P (1993) Biol Mass Spectrom 22:338
34. Pappin DJC, Højrup P, Bleasby AJ (1993) Curr Biol 3:327
35. James PQ, Carafoli E, Gonnet G (1993) Biochem Biophys Res Commun 195:58
36. Yates JR, Speicher S, Griffin PR, Hunkapiller T (1993) Anal Biochem 214:397
37. Shevchenko A, Wilm M, Vorm O, Mann M (1996) Anal Chem 68:850-858
38. Cahill DJ, Nordhoff E, O'Brien J, Klose J, Eickhoff H, Lehrach H (2001) In: Pennington SR, Dunn MJ (eds) Proteomics. From protein sequence to function. BIOS Scientific Publishers, Oxford, p 1
39. Schmidt F, Lueking A, Nordhoff E, Gobom J, Klose J, Seitz H, Egelhofer V, Eickhoff H, Lehrach H, Cahill DJ (2002) Electrophoresis (in press)

Received: April 2002

Adv Biochem Engin/Biotechnol (2003) 83: 189–209
DOI 10.1007/b11117

Topological Proteomics, Toponomics, MELK-Technology

Walter Schubert

MelTec Ltd., ZENIT-Building, Leipziger Strasse 44, 39120 Magdeburg, Germany
E-mail: info@meltec.de
Institute of Medical Neurobiology, Otto von Guericke-University of Magdeburg,
Leipziger Str. 44, 39120 Magdeburg, Germany

MELK is an ultrasensitive topological proteomics technology analysing proteins on the single cell level (Multi-Epitope-Ligand-'Kartographie'). It can trace out large scale protein patterns with subcellular resolution, mapping the topological position of many proteins simultaneously in a cell. Thereby, it addresses higher level order in a proteome, referred to as the toponome, coding cell functions by topologically and timely determined webs of interacting proteins. The resulting cellular protein maps provide new structures in the proteome: single combinatorial protein patterns (s-CPP), and combinatorial protein pattern motifs (CPP-motifs), bound to superior units. They are images of functional protein networks, which are specific signatures of tissues, cell types, cell states and diseases. The technology unravels hierarchies of proteins related to particular cell functions or dysfunctions, thus identifying and prioritising key proteins within cell and tissue protein networks. Interlocking MELK with the drug screening machinery provides new clues related to the selection of target proteins, and functionally relevant hits and drug leads. The present chapter summarizes the steps that have contributed to the establishment of the technology.

Keywords. MELK, Whole cell fingerprinting, Topological proteomics, Toponomics, TOPONOME, Functional proteomics

1 Introduction

The knowledge of the proteome – the snapshot of the total protein output encoded by a genome – provides important information on: (a) if and which genes are translated, (b) the relative abundance of expressed proteins and (c) the post-translational modification of these proteins. Large scale protein profiling techniques addressing these issues are constructed on the basis of a large number of cells and tissue homogenates. On the basis of proteins, which are extracted from cell homogenates these techniques provide insight into the protein systems 'working' in a large number of cells and tissues at one time. Evidence indicates that these techniques are particularly valuable for the identification of proteins, which are specifically regulated under defined conditions, such as in disease. Whilst we are now ready to understand the nature of proteomes on the level of these analytical proteomic technologies, it is clear that these techniques provide only a partial view of the functionalities in a proteome. We still face a significant next problem to be solved: how are proteomes organized in a cell or a tissue? This question cannot be answered when cells are destroyed by homogenisation procedures, because the 3D information on the cellular distribution of the proteins within a proteome are lost. Further – if we were able to quantify the organization of proteomes, or at least large fractions of proteomes in individual cells (i. e. many protein families in parallel) – is there any significant additional information related for example to the identification of target proteins?

It is conceivable that the organization of complex unicellular and higher organisms involves the proper spatial distribution of macromolecules, metabolites, ions and cell organelles, the interplay of these components, and their individual regulation due to functional requirements. The myriads of different cellular functions are the result of defined webs of interacting proteins (functional protein networks) which are determined in time and space in every cell. Consequently, every protein having a *molecular function* by its biochemical properties (i. e. well defined domains etc.) exerts *cellular function(s)* by interaction with other proteins, i. e. as an element within one or multiple different protein networks (Fig. 1).

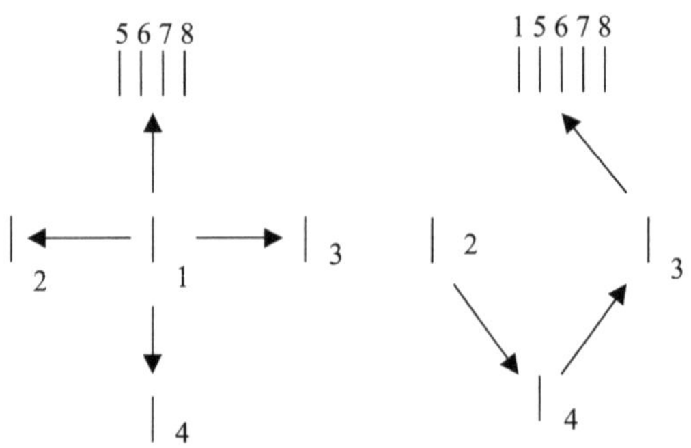

Fig. 1. Different protein networks are determined by the differential spatial location of their elements (proteins). Translocation of protein 1 to protein complex (5, 6, 7, 8) leads to a new protein network

Within these networks every protein is determined topologically, i. e. has to be at the right place at the right time to interact with other proteins, the latter being also defined by individual locations at the required time. How and when these networks are formed is decided by every cell, more or less under control of other cells, i. e. cell-to-cell or cell-to-matrix interactions in tissues. Detection and quantitation of cellular protein networks addresses higher level order functions in cellular proteomes. It can uncover directly the protein hierarchy within these interacting protein webs. To identify the myriads of different protein networks in a proteome is an important key to understand how a proteome encodes the many cell functions and behaviours, like cell polarization, migration, invasion, intercellular communication during morphogenesis etc. 'Reading' and translation of this information into drug discovery machineries requires technologies for the simultaneous topological analysis of a large number of different proteins in morphologically intact cells and tissues.

The present chapter summarizes progress in identifying cellular patterns of many proteins in parallel on the single cell level. The underlying technology, MELK, can trace out directly the relative location of tens or hundreds of different proteins and other molecular classes in thousands of cells simultaneously. Principally, any complex protein pattern identified by this approach on the submicrometer to micrometer scale of a cell is a three-dimensional (3D) protein pattern that can be quantified by specific algorithms in different ways. The latter may address the identity and quantity of the proteins within many subcellular volumes or the cell as a whole single "unit", in which all identified proteins are mapped as combinatorial binary codes. This latter approach may be of great value for fast searches for differences between a normal and a pathological tissue. Cells may be cultured, or isolated from body fluids, or present in tissue sections obtained from biopsy material. The prospect of this 'single cell 3D-proteomics' technology is to analyse functional protein networks, which cannot be detected in large scale expression profiles based upon tissue homogenates.

2 Technological Principle

To map tens or hundreds of different species or classes of molecules in morphologically intact cells a robotic imaging technology, referred to as MELK (multi-epitope-ligand-"Kartograph") was established [1]. The technology, based upon a special fixation of cells or tissues, uses large molecular libraries (i. e. antibodies, peptides, random proteins, lectines or any ligands) to detect and localize the many individual molecular species in cells by means of photonic signals, i. e. fluorescence. It is based on a biophysical principle of diffusion kinetics within defined subcellular volumes (successive protein ligand aggregation stable diffusion channelling, SPLASDIC) (to be published elsewhere). Each individual molecule in a cell is detected and registered as a spatial signal map, and aligned relative to other molecular signals in the same cell. MELK is a fully automated high throughput technology using multiple integrated pattern recognition and data mining algorithms to identify protein networks, which are specific or selective for a cell type or a cell state in health and disease. An integrated system of algorithms

Table 1. Synopsis of the MELK-technology

Molecular classes in a cell	Technology MELK	Tools	Detection principle	Results	Cellular organisation level
proteins/peptides → carbohydrates → nucleic acids → lipids →	3D localisation in parallel in single cells	- protein libraries, i.e. antibodies - lectin libraries, - ligand libraries, I.e. nucleid acids, small molecules, others - random naïve libraries	SPLASDIC „successive protein ligand aggregation stable diffusion channeling"	molecular networks	TOPONOME

learns to interpret the patterns by data matching and identifies and prioritises target proteins directly from original or matched data. Basically, the technology is a biology-driven pattern recognition system (to be published in detail elsewhere).

MELK analyses the topological order, i.e. the cellular organization, of the major molecular classes of a cell – proteins, carbohydrates, lipids, nucleic acids. It therefore addresses the organizational equivalent of genome and proteome in a cell, referred to as a toponome[1] (Table 1). Consequently, a toponome contains all functional protein networks of a cell.

The following biological examples may explain the working of the technology.

3 A Disease Example

It is a daily routine procedure in clinical medicine to substantiate a diagnosis by means of tissue biopsies. These clinical tissue samples (i.e. diagnostic biopsy material) with clearly distinguishable tissue compartments, i.e. histologically defined tissue structures, were found to be ideal to develop a topological approach to protein-systems in cells [2]. In particular muscle tissue samples of neuromuscular disorders, having important impact on research in related fields [3–14], provided histologic criteria.

Mammalian skeletal muscle tissue is composed of structurally well defined skeletal muscle fibres, connective tissue bands separating individual fibres from each other, and a dense capillary network surrounding each muscle fibre. Individual muscle fibres are ensheathed by a basal lamina cylinder. Under physiological conditions small mononuclear satellite cells, which constitute an important source for muscle regeneration, are present [5]. However, under specific inflammatory disease conditions of polymyositis, a chronic autoimmune disease of the human T lymphocytes penetrate small blood vessels and first accumulate in the perivascular tissue compartments [8]. Cells then develop an enormous

[1] The term toponome is composed of the Greek nouns topos (place, position, region) and nomos (order, law).

migratory potential. They start to migrate into the connective tissue space between muscle fibres (endomysium) and select one individual muscle fibre to penetrate the basal lamina cylinder. Interestingly, these cells ignore other basal lamina cylinders of neighbouring fibres. Figure 2 illustrates schematically this process of T cell invasion, showing three characteristic steps. Stage 1 shows the 'pace maker' T cell penetrating a basal lamina cylinder of a muscle fibre. In stage 2, the initially invasive pace maker T cell has displaced the plasma membrane of the muscle fibre and has been followed by several other T cells behind the invasive front. In stage 3, successive accumulation of a large number of invasive T cells inside the basal lamina cylinder has led to an almost complete compression and displacement of the muscle fibre. This compression finally leads to mechanical rupture of the muscle fibre and loss of function.

A major goal is to understand the differential adhesive mechanisms at the T cell surface, which allow the T cells to organize themselves as an abnormal,

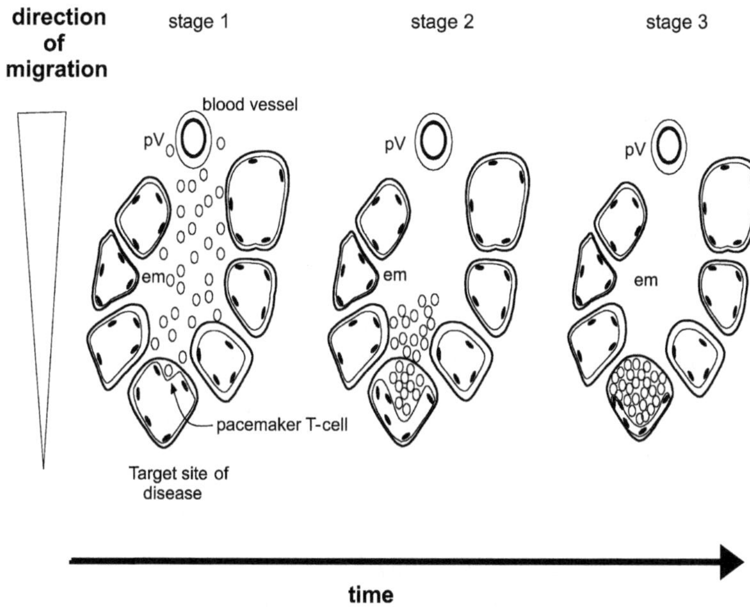

Fig. 2. Schematic illustration of three characteristic stages of muscle invasion of T lymphocytes in a human chronic inflammatory muscle disease (polymyositis). Stage 1: after extravasation from the blood T lymphocytes have formed an abnormal coherent lymphoid tissue extending across different muscle tissue compartments: the perivascular (pv), and the endomysial (em) territories. One single T cell (the invasive front T cell, or pacemaker T cell) has penetrated the basal lamina cylinder and started to displace the plasma membrane of a muscle fibre (*arrow*). This latter site is the target site of this disease. Note that all cross-sectioned multinucleated muscle fibres are ensheathed by a basal lamina cylinder, BLC (*dark black line*). Several T lymphocytes, which are assembled behind the invasive front, are in close contact with the BL and ready to follow the pacemaker T cell. Stage 2: progressed invasion of the BLC by T cells has led to partial compression of the muscle fibre. Stage 3: T cell invasion is completed. All T cells are inside the BLC. The muscle fibre is mechanically compressed and shows a small rim, ready to rupture

aberrant lymphoid tissue compartment to select individual muscle fibres and finally drive the invasion towards the basal lamina cylinder. Important new clues to these mechanisms are obtained, when 17 different monoclonal antibodies against known cell surface receptor proteins with adhesive properties are used to map these proteins on the cell surface of invasive T cells [15] (Table 2).

These randomly selected 17 proteins, like any other proteins, do not distribute randomly on the cell surface. Rather, the level of each protein depends on the biological state of the individual cell and on the local environmental cues. The most simple case is that a protein, i. e. CD38, is present on the surface of all cells, or completely absent. However this is not the case. MELK analyses revealed that this protein is only present in a given tissue microenvironment. If these MELK analyses were performed for the simultaneous localization of 17 different cell surface proteins, it became obvious, that the CD38 protein occurs in specific combinations with 7 different other proteins out of 17 (Table 2). Equally, specific combinations were found of selected proteins in other microenvironments inside the lymphoid tissue. Table 2 indicates these specific combinations on the level of individual T cells at defined sites of stage 1 of the T cell invasion (compare with Fig. 2). By this example it can be shown that, on the basis of parallel localization of a limited number of individual proteins on the single cell level, distinct stages of the invasion process can be identified. Consequently, many data sets of this kind, covering all proteins of a cell or a tissue, would be the MELK equivalent to data sets like genome or proteome. This data set is referred to as a protein TOPONOME. The smallest subunit in a protein Toponome is the single combinatorial protein pattern, briefly s-CPP (Table 2, horizontal lines: annotations 1 to 30). In the present example this would be equivalent to a combinatorial protein pattern of a single cell: each protein of an s-CPP is registered as a binary code per cell, i. e. present or absent above a threshold level [1/0] (Table 2). A specific set of s-CPP which is characteristic for an individual biological state or function is referred to as combinatorial protein pattern motif, or simply CPP-motif. Table 3, which is a simplified representation of Table 2, illustrates the CPP-motifs which are associated with different T cells during the invasion across the muscle tissue. Principally, s-CPP and CPP-motifs can be found by MELK analyses at the cellular (see Tables 2 and 3) and the subcellular level (not shown). On the subcellular level, s-CPP and CPP-motifs are associated with the different subcellular compartments, i. e. organelles or membranes.

Within a CPP-motif single protein species may be identified, which are present in all s-CPP belonging to the corresponding CPP-motif. These proteins are termed *lead proteins*. As outlined in Table 3, CD38 is a lead protein of motif 2 within stage 1 of the T lymphocyte invasion process in the muscle tissue. Given the MELK analyses shown in Table 3, four different CPP-motifs can be distinguished, which are unique to the T cells moving from the perivascular environment across the endomysial environment towards the basal lamina cylinder of a muscle fibre. Hence this whole T cell translocation or migration process can be described as a follow-up of defined s-CPP on the T cell surface until the T cell reaches the basal lamina cylinder. During migration, the trans-

Table 2. Binary codes of cell surface proteome fractions, which are characteristic for T cell invasion in polymyositis (15). Additional binary codes specific for non-T cells (endothelial cells, muscle cells) in muscle tissue are also shown (fractions 6 to 8). These binary codes were obtained by simultaneous MELK-mapping of 17 different protein epitopes and ligand binding sites for carbohydrate structure, respectively (specification in the upper horizontal line). Presence or absence of these epitopes or ligand binding sites (vertical lines) were mapped as binary codes [1/0] above a fluorescence threshold level. Each horizontal line gives the single combinatorial protein pattern (s-CCP) binary code per single cell. Each s-CCP may be present in at least one cell. Each s-CCP is mapped to its topological site in the tissue: pv: perivascular accumulation of T cells (perivascular lymphoid tissue); pre-em: transitional area between perivascular and endomysial tissue; em: accumulation of T cells within the endomysium (connective tissue between muscle fibres), in which T cells migrate towards a target muscle fibre; pre-in: T cells contacting the basal lamina of individual muscle fibre; in: penetration of the basal lamina of individual muscle fibre by pacemaker T cells, thereby invading the basal lamina cylinder. Note that the arrow denotes the invasion pathway of T cells entering the muscle tissue from the blood circulation (compare with Fig. 2)

proteins / glyco-structures		APP	CD 16	CD 71	CD 11b	CD38	HLA-DR	HLA-DQ	CD 2	CD 7	CD 3	CD 8	CD 4	CD 45 RA	CD 36	CD 56	CD 62 L	CD 57	Lami-nin A	UEA I lectin
pv	1	0	0	0	0	0	0	0	0	0	0	1	0	0	0	0	0	0	0	0
	2	0	0	0	0	0	0	0	1	1	1	1	0	0	0	0	0	0	0	0
	3	0	0	0	0	0	0	0	1	0	1	0	1	0	0	0	0	0	0	0
	4	0	0	0	0	0	0	0	1	0	0	1	0	0	0	0	0	0	0	0
	5	0	0	0	0	0	0	0	0	0	0	0	1	0	0	0	0	0	0	0
pv	6	0	0	0	0	1	0	0	0	1	1	1	0	0	0	0	0	0	0	0
	7	0	0	0	0	1	0	0	0	1	1	0	0	0	0	0	0	0	0	0
	8	0	0	0	0	1	1	1	0	0	1	0	0	0	0	0	0	0	0	0
	9	0	0	0	0	1	0	0	0	1	0	1	0	0	0	0	0	0	0	0
	10	0	0	0	0	1	1	0	1	1	1	1	0	0	0	0	0	0	0	0
	11	0	0	0	0	0	0	0	1	0	1	1	0	0	0	0	0	0	0	0
em	12	0	0	0	0	0	1	0	0	0	0	0	0	0	0	0	0	0	0	0
	13	0	0	0	0	0	1	0	0	0	0	1	0	0	0	0	0	0	0	0
	14	0	0	0	0	0	1	1	0	1	0	0	0	0	0	0	0	0	0	0
	15	0	0	0	0	1	0	0	0	0	0	0	0	0	0	0	0	0	0	0
pre-in	16	0	0	0	0	0	0	0	0	0	1	1	0	0	0	0	0	0	0	0
	17	0	0	0	0	0	0	0	0	1	0	1	0	0	0	0	0	0	0	0
in	18	1	0	0	0	0	0	0	0	0	0	1	0	0	0	0	0	0	0	0
mature muscle fibres	19	0	0	0	0	0	0	0	0	0	0	0	0	0	1	0	1	0	1	1
	20	0	0	0	0	0	0	0	0	0	0	0	0	0	0	0	1	0	1	0
	21	0	0	0	0	0	0	0	0	0	0	0	0	0	0	1	1	0	1	1
	22	0	0	0	0	0	0	0	0	0	0	0	0	0	0	0	1	0	1	1
developing myotubes	23	1	0	0	0	0	0	0	0	0	0	0	0	0	1	1	1	0	1	1
	24	1	0	0	0	0	0	0	0	0	0	0	0	0	0	1	0	0	1	0
(regeneration)																				
endothelial cells	25	1	0	0	0	0	0	0	0	0	0	0	0	1	1	0	0	0	1	0
	26	1	0	0	0	0	0	0	0	0	0	0	0	0	1	0	0	0	1	1
	27	1	0	0	0	0	0	0	0	0	0	0	0	0	1	0	0	0	1	0
myogenic stem cell	28	1	0	0	0	0	0	0	0	0	0	0	0	0	1	0	0	0	1	0
	29	1	0	0	0	0	0	0	0	0	0	0	0	0	1	1	0	0	1	1
	30	1	0	0	0	0	0	0	0	0	0	0	0	0	0	1	0	0	1	0

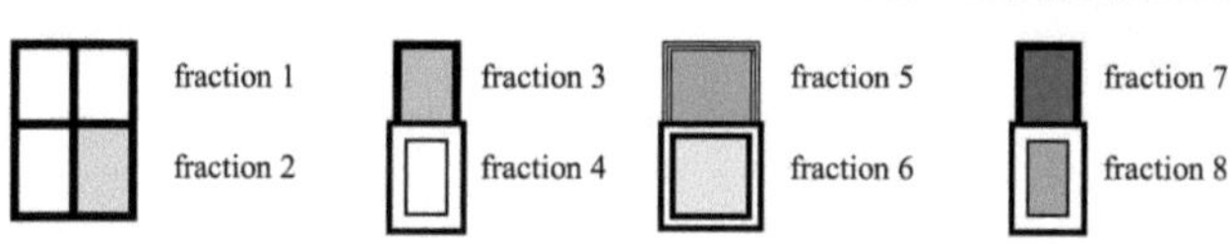

Table 3. Illustration of combinatorial protein pattern motifs (CPP-motifs), present on the surface of T cells during invasion of muscle tissue. The binary codes shown here are depicted from Table 2 to illustrate the formal characteristics of CPP-motifs and of transitional elements (TE) (for details see text). Note the significance of CPP-motifs as signatures for specific territories in the lymphoid tissue. The large arrow indicates the direction of T cell invasion from the perivascular territory to the target site of muscle fibre invasion (compare with Fig. 2, stage 1). Small arrows indicate the *leading proteins* of the different CPP-motifs. The strong line denotes the basal lamina, the thin line denotes the plasma membrane of the muscle fibre located inside the basal lamina cylinder. One T cell, denoted as TE5 (the pacemaker T cell) has penetrated the basal lamina cylinder. It is located between the basal lamina and the plasma membrane of the muscle fibre

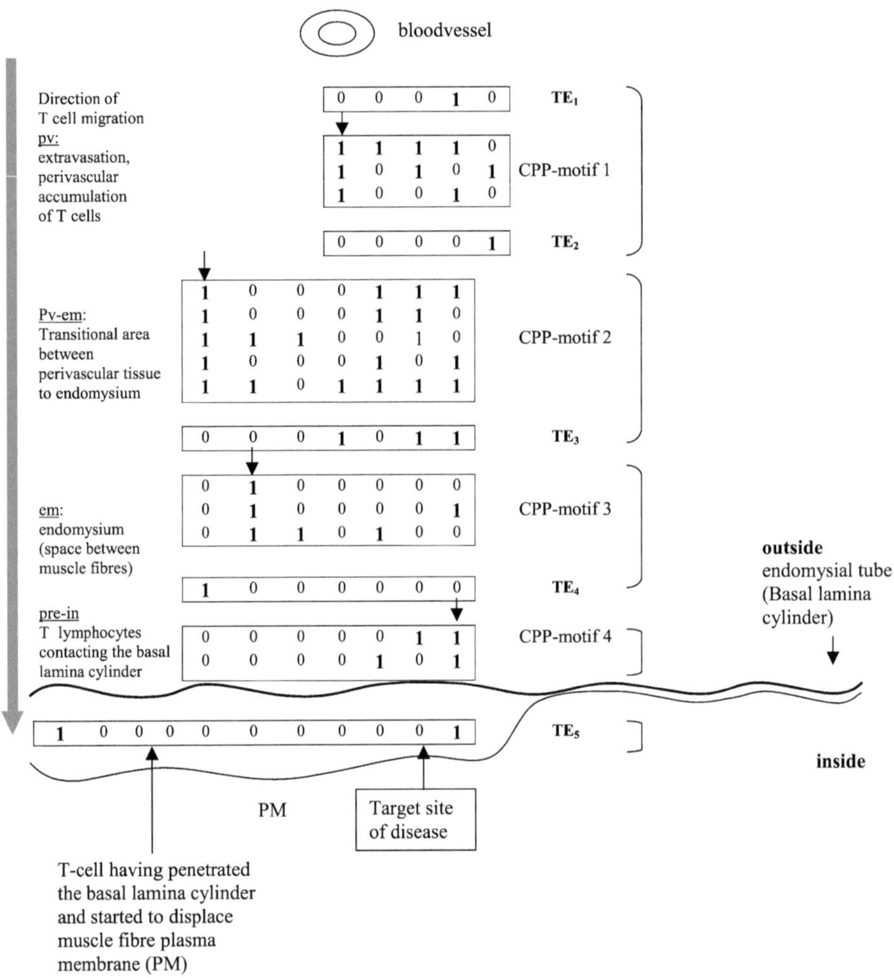

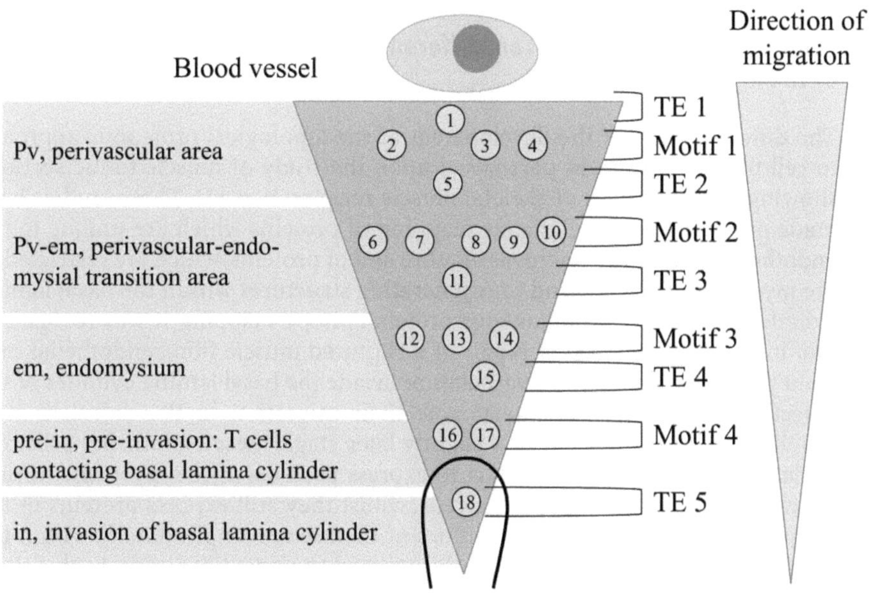

Fig. 3. Summary of Tables 2 and 3, illustrating the synopsis of s-CPP (annotations 1 to 18), CPP-motifs and TE on the cell surface of invasive T cells, in the different territories of the lymphoid tissue structures in polymyositis

location of a T cell from one microenvironment to the next, i.e. from the perivascular area to the endomysium, is characterized by a particular s-CPP on the cell surface, termed transitional element (TE). TEs are intermingled between two characteristic CPP motifs. This is illustrated schematically in Fig. 3. In the present example, TEs obviously represent a functional state of a migratory T cell, in which the T cell surface adapts to a new microenvironment. Interestingly during this process of migration the T cell appears to acquire selectivity for penetration of the basal lamina cylinder by continuous reduction of the number of proteins within the s-CPP on the cell surface: from seven proteins in CPP-motif 2 down to only two proteins in TE5 (Table 3). The latter is a specific assembly of the CD8 receptor protein and the Alzheimer amyloid precursor protein (βAPP). βAPP/CD8 assemblies on the surface of T cells lacking the other 15 proteins, which were mapped simultaneously by MELK (TE5, Table 3) is highly non-random, and in this constellation was only observed in polymyositis [15].

This observation raises the possibility that cellular 'autoimmunity' may occur, when T lymphocytes generate abnormal s-CPP and CPP-motifs on the cell surface to build up pathologic cell-to-cell interactions. Furthermore, MELK data suggest that organ-specific T lymphocyte homing in autoimmune disorders may be the result of abnormal CPP-motifs on the cell surface of circulating T lymphocytes (to be published).

4
Topological Detection of Transdifferentiation: the Riddle of the Myogenic Stem Cell

The different steps of the development of the topological proteomic approach to cell functions were in part based upon the study of muscle tissue sections showing various stages of skeletal muscle regeneration [2]. These studies have made possible the simultaneous mapping of proteins which are unique to the endothelial cells of the microvasculature and of proteins, which are restricted to the myogenic stem cells and to regenerating structures within the basal lamina cylinder [2]. By this simultaneous protein epitope mapping it was recognized that in a very early stage of repair of a ruptured muscle fibre, endothelial cells from the surrounding microvasculature invade the basal lamina cylinder of the defective muscle fibre and terminate their migration in the defective zone (annotation 28 in Table 2). In a slightly later stage, these endothelial cells dissociate from each other and start to express proteins of the myogenic lineage characteristic for myogenic stem cells, whilst they still express proteins of the endothelial cell lineage. Cells then orient themselves longitudinally within the basal lamina cylinder and are closely attached to intact fibre ends. In this stage these cells continuously downregulate the endothelial marker proteins, whilst they upregulate strongly the NCAM proteins which are characteristic for mononuclear myogenic stem cells (annotation 30 in Table 2). Cells then proliferate and fuse to form myotubes, bridging and finally repairing the defective zone. MELK-data sets revealed that the latter are characterised by unique CPP-motifs (annotations 23 and 24 in Table 2) [2].

Hence, on the basis of the relative localisation of seven reference proteins, evidence was provided for the presence of transdifferentiation of local endothelial cells as a mechanism to form myogenic stem cells in human muscle tissue. The classical view of muscle fibre repair however was that satellite cells, which are of embryonic origin and reside within the basal lamina cylinder, are the source for muscle fibre repair [16].

Consequently, these topological data indicating transdifferentiation of local endothelial cells were strictly against the grain of what we think we know about fundamental aspects of muscle biology. They stimulated the discussion on the source of myogenic stem cells and the nature of the satellite cells (i.e. communication, controversial discussion by G. Cossu during the annual conference of the German society of Cell biology, Konstanz, Germany, 1992). These data also provided the fundamentally new concept that adult, mature cells can transdifferentiate to a stem cell, when they receive the correct environmental cues.

Recently, by a combination of embryological, cell biological and transgenic technologies, the Cossu group confirmed experimentally that satellite cells may derive from the vasculature system [17].

The present example strongly validated the topological principle (proof of concept), that the simultaneous localisation of reference proteins "superimposed" on known structures in a tissue can uncover unknown cellular mechanisms.

Furthermore, the example may suggest that the simultaneous localization of hundreds of different proteins by MELK will allow one to address transdifferentiation mechanisms in any adult organ, because it allows one to trace out cell type specific programs of mature cells and of stem cells simultaneously.

5
Why Explore Protein Networks?

Theoretical considerations indicate that recognition of the range from protein structure to molecular and cellular protein function requires a synopsis between large scale expression profiling, structural proteomic and topological proteomic technologies. Each of these levels of recognition has its own rules and indications. This may be illustrated by the following example.

Given a normal and an abnormal cell both expressing four proteins at identical abundances (Fig. 4), the normal cell exhibits assemblies of these proteins, above defined as s-CPP (see Chap. 3), restricted to the cell surface. In the abnormal cell, one protein is stored inside the cell and cannot reach the cell surface. The resulting new s-CPP on the cell surface may exert new cell-to-cell interactions, which might be causing disease simply by the abnormal combination of otherwise normal proteins. New disease specific s-CPP of this kind is considered as a "tag" of a more extended pathogenic protein network (see below). This network may operate, for example, in a few tumour cells, a condition which may lead to invasion and formation of metastases.

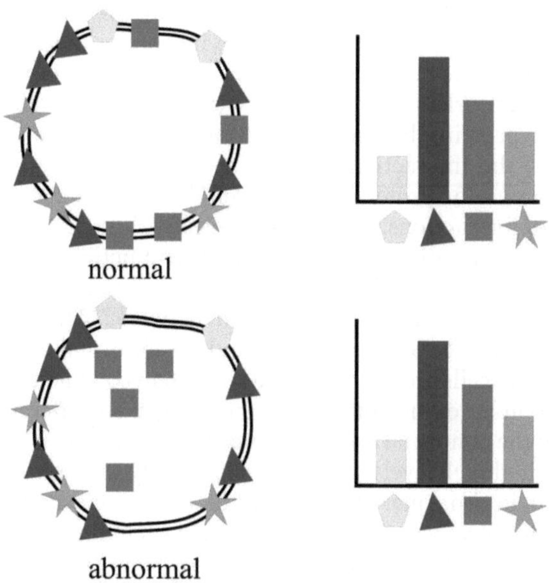

Fig. 4. Schematic illustration comparing cellular protein patterns traced out by MELK (a subfraction of the cellular proteome, *left hand side*) with protein profiles (*right hand side*) as provided by large scale proteome profiling techniques based upon tissue homogenates. Note that the differences between the normal and the abnormal cell are undetectable by expression profiling

Since only the spatial protein distribution, but not the level of the individual protein, is altered, homogenization procedures of these cells in order to extract and biochemically identify and quantify the proteins leads to identical protein profiles of the normal and the abnormal cells. In the present example the interesting target proteins are among the normal cell surface proteins forming abnormal patterns, which would not be conspicuous in large scale protein analyses (Fig. 4, right hand side). Strategically, once the abnormal patterns have been identified by MELK proteomic mapping, a procedure of 'downstream' analyses will follow involving biochemical, structural and functional analysis of the identified protein pattern/complex. Hence, the MELK pattern on analysis leads to the identification of relevant functionally linked proteins or protein assemblies within a cells toponome and thereby sets a window in the mass of proteins expressed in a tissue.

How can the topological information be quantified? Given two parameters: (a) the identity, and (b) the abundance of a protein. Given further that each protein can be characterized, i.e. in a 2D Gel, at 50 to 100 abundance levels, and that 20 proteins are identified in a 2D gel on the basis of homogenization of one million cells. The maximum of theoretical protein combinations related to the above parameters will then correspond to an information of 100^{20} bits, at best. Given further that a cell-wise measuring of these 20 proteins simultaneously by MELK would allow one to quantify these profiles within approximately 2000 different subcellular volumes in every cell, then the resulting maximum of information related to one million cells is $2 \times 10^9 \times 100^{20}$ bits. The resulting gain of information is 2×10^9 bits. Since the latter number reflects the cell as the topological, organisational unit of the proteome, it contains the relevant information on functional linkages of proteins, i.e. the functional protein networks. A functional protein network is considered to be (i) a specific protein complex, (ii) the interaction of proteins by means of diffusible molecules (i.e. substrates of enzymes), (iii) the interaction of proteins by directed transport of proteins, protein-substrate – or protein-protein-complexes or (iv). a combination of all these modes of interactions. This is schematically shown in Fig. 1, where protein 1 is an element of two different functional protein networks.

What is the interrelationship between combinatorial protein patterns (s-CPP or CPP-motifs) and a functional protein network? This may be explained schematically in Fig. 5. Given a normal and an abnormal cell, both the normal and the abnormal cells are characterized by protein complexes at the cell surface. The protein complex of the normal cell consisting of six different proteins (7 – 12) is modified in the abnormal cell by absence of protein 8. This can be detected directly on the single cell level by MELK. Functionally, this abnormal protein complex may result in abnormal cell-to-cell interaction. For example, in tumour cells this may lead to selective organ-invasion, when such tumour cells, circulating in the blood, encounter the cell surface of organ-specific endothelium and *adhere* to it selectively by the abnormal protein complex. Since it is likely, that the abnormal protein complex is interrelated to intracellular regulatory protein networks operating in different compartments of the cell, the abnormal cell surface protein complex, once detected as specific, can be regarded as a tag to search for other parts of the protein network associated with this tag on the single cell level.

$s\text{-}CPP_1=(7,8,9,10,11,12)$; $s\text{-}CPP_2=(1,2,3,4,5,6)$;

$s\text{-}CPP_3=(7, 9,10,11,12)$; $s\text{-}CPP_4=(13,14,4,15,16,17,18)$

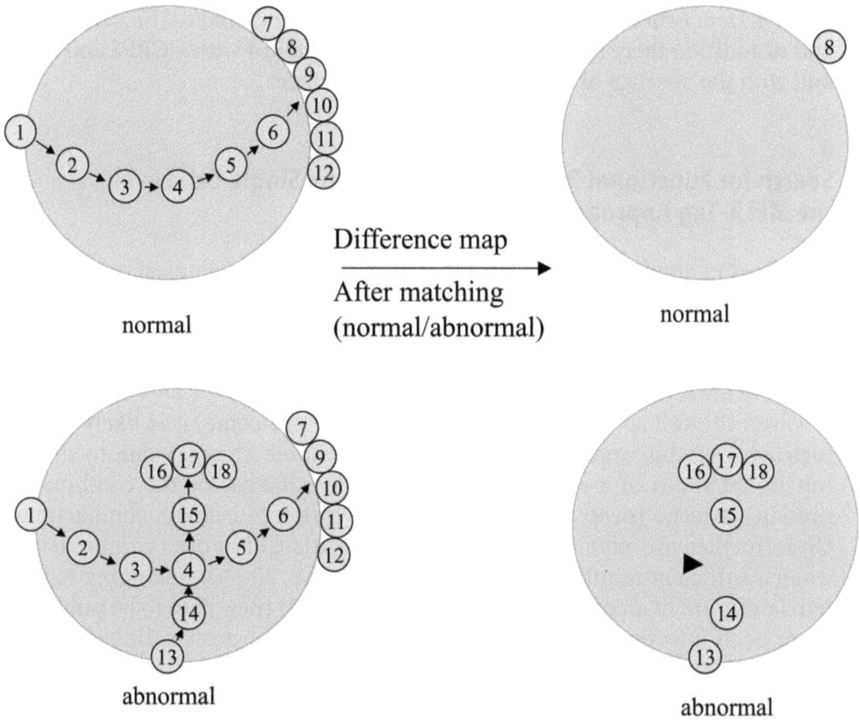

Fig. 5. Schematic representation of a *normal* and an *abnormal* cell. The normal cell is characterized by the interaction of the proteins (1 → 6) with a cell surface protein complex (7 → 12); the abnormal cell is characterized by an additional protein network (4, 15 – 18) and absence of protein 8 in the cell surface protein complex. The difference map, obtained by matching, reveals a specific linkage of protein 4 to two different protein networks (*arrowhead*) and to the absence protein 8 in the abnormal cell. The different parts of the present protein networks are detected as s-CPP by MELK

Simultaneously with this tag, intracellular proteins are systematically screened in an organelle-wise and a protein-family-wise manner. For example, proteins belonging to signal cascades, enzyme families, etc. or proteins, which can be regarded as reference proteins for subcellular organelles, are screened in parallel in the same cell. Each of these families may show new s-CPP in the intracellular compartments of the abnormal cell. By matching MELK data sets of the normal and the abnormal cells, subcellular s-CPP, which are unique to the abnormal cell may be identified. The latter are new tags, which then indicate the compartments or the family of proteins, which are specifically linked to the abnormal cell-surface protein complex, or tag (Fig. 5, s-CPP2). Such a new tag may be the starting point for a further directed MELK screening for proteins which are specifically associated with the corresponding compartment. One example, outlined in Fig. 5, shows that by this approach one new subcellular s-CPP (Fig. 5, s-CPP1) is found that is present in both the normal and the abnormal cell, whilst one additional s-CPP is unique to the abnormal cell (Fig. 5, s-CPP4). The example also indicates presence of a protein (Fig. 5, protein 4) which is an element both of s-CPP2 and s-CPP4. Hence there is a functional linkage of the abnormal s-CPP3 and s-CPP4, and in addition there is a functional linkage of protein 4 with s-CPP2 and s-CPP4 and with the absence of protein 8 on the cell surface.

6
Search for Functional Protein Linkages on the Single Cell Level: the MELK-Tag Approach

The above examples are based on experiences with the topological proteomic approach in selected fields [2, 15, 18]. Although much remains to be learned in the future using MELK technology, the present knowledge on the selectivity of s-CPP may allow some generalisations on the use of MELK to identify functional protein networks.

Given the cell as the organisational unit of a proteome, it is likely that disturbances of this organisation, or specific dynamic changes due to different functional states of a cell, may be reflected by changes of the combinatorial protein patterns (protein assembles) in the different cellular compartments. Given further the experience that disease-specific CPP-motifs can be detected when a sufficient number of different proteins (i.e. 20–40) belonging to a protein family are mapped simultaneously by MELK [2] (new data to be published), there is an effective strategy to search for functional protein linkages on the single cell level.

At first, antibody libraries against many different protein families are constructed on the basis of previous MELK-data sets or on the basis of protein specific sequence information derived from the genome: for example 40 antibodies against adhesion proteins of the cell surface, 40 antibodies against cell surface channel proteins, 40 antibodies against growth factor receptors, 60 antibodies against reference proteins for diverse subcellular compartment, etc. Based on previous MELK-data sets it is held that these antibody libraries must not be complete to find selective patterns. In a next step each of these libraries is used to search for s-CPP or CPP-motifs which are characteristic for a given biological state of a cell or a cell type. Each individual approach in topological proteomics, i.e. the sequence of steps towards functional protein linkages, is based upon cell biological considerations: for example, invasive tumour cells are first screened for their adhesive cell surface proteome fraction, because abnormal invasion is most likely due to abnormal adhesive functions, etc. Opposed to that, metabolically altered liver cells will be first screened with MELK antibody sets against subcellular enzyme systems, and so forth.

Once an s-CPP or CPP-motif characteristic for a given biological state has been found by one of the above "search libraries" the latter is classified as a tag, which now allows for in-depth search for proteins which are specifically linked to it. This in-depth search may be performed in several ways: (i) the MELK screening is extremely extended by using antibody libraries against other protein families likely to be present in the same compartment of the detected tag; (ii) the MELK screening is extended for search-libraries specific for proteins present in subcellular compartments other than that of the identified tag; (iii) the MELK screening is extended for search libraries specific for proteins, the compartmentalisation of which within the cell is unknown. Together the advantage of this MELK-tag approach is that it rapidly leads to the detection both of a relevant protein family and the associated compartment as a first tag for in-depth analyses. Furthermore MELK can be correlated with 2D-gel based large scale protein profiling techniques allowing one to optimise the identification of proteins linked to disease processes. Finally, it is also feasible to use naive antibody libraries to screen for s-CPP or CPP-motifs. This approach may be particularly suited to identify, among a large number of naïve antibodies, relevant antibodies to be used to isolate the underlying proteins recognised as being specifically linked to disease.

It is held that all tags which, by definition are found to be specific for a biological state of a cell, are functional in the sense that they constitute part of a functional protein network, which unequivocally "enciphers" the corresponding biological state. In many cases, for example in cells in vitro, this can be tested experimentally. Details are beyond the scope of the present chapter.

7 Automatic Recognition of Cell Structures In Situ

The first step in a topological proteomics approach, as outlined above, is high-throughput MELK reading of cells using the SPLASDIC detection principle, in tissue sections [1, 2]. In a next step, detection, quantitation, matching and classification of a large number of cells related to their expressed proteome fractions, their topology and their relative locations in a tissue according to the above described elementary units (s-CPP, CPP-motifs) is performed. We have developed several algorithms for the automatic detection of any random structure in situ, one of which is a general pixel/voxel algorithm for the quantitation of proteomic patterns within a given volume of a cell (i.e. $60 \times 60 \times 60$ nm) (to be published elsewhere). In addition there is a more specified algorithm adapted, for example, to the high-throughput measuring of mononuclear cells, i.e. lymphocytes, in tissue sections [19–24].

Detection algorithms related to automatic cell detection, which were reported hitherto, are based on the idea of fitting a model to a gradient ensemble [25, 26], wave propagation [27], boundary tracing [28] or use of Hough Transformation [29, 30] to detect circle-like objects. Cells in tissues show however considerable variation of their shapes and sizes due to migration, cell to cell interaction, tissue-specific constraints and so on. These variations render it difficult to define a model fitting for all possible cell types. To address these high variables we have chosen migratory lymphocytes as a biologically relevant example to construct a neural cell detection system (NCDS), which uses an adaptive neural classifier to map each image point **P** in a fluorescence micrograph to an evidence value $C(p). \in [0;1]$. The magnitude of the evidence value estimates the probability

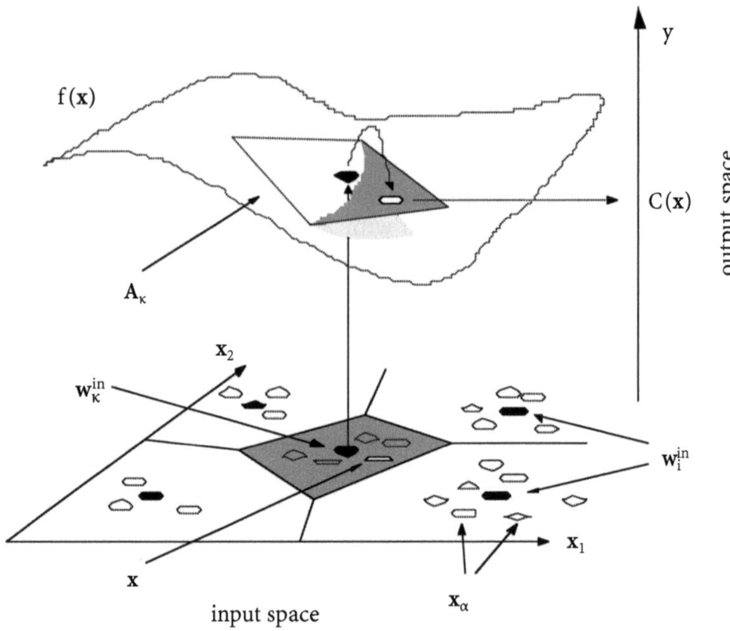

Fig. 6. Example illustrating a Local Linear Map approximating a function $f(x), x \in \Re^2$ (*dotted plane*) with five nodes, LLM's nodes' w_i^{in} form Voronoi cells of the input space. The plane is approximated by projections of the cells (one sketched as *dark grey plane*). An input vector x is mapped to $C(x)$ in the output space. First, the nearest neighbour w_κ^{in} to the input is selected, then the input is mapped via the coupled matrix A_k (*dashed arrows*)

for the image point to be occupied by a fluorescent cell. A local maximum search in the evidences of all image points combined with a thresholding procedure then can deliver the described positions of fluorescent cells. This classifier is a particular species of an artificial neural net which has turned out to be very efficient in fast learning of non-linear classification tasks from even small training sets.

Figure 6 illustrates the functionalities of the NCDS, referred to as Local Linear Map (LLM). To analyse the stabilities and reproducibility of the LLM results we have compared detection of lymphocytes in inflammatory muscle tissue with lymphocytes in the tonsil, which is a tissue containing lymphocytes in the form of densely packed cellular nets, a condition rendering it extremely difficult to distinguish cellular boundaries [19]. Figure 7 illustrates a low magnification of fluorescent lymphocytes in muscle tissue and in a tonsil together with the output images of the LLM. These analyses have shown that the NCDS recognizes at least 95% of the cells correctly and can trace out 40,000 cells within 3 h. The prerequisite for this approach is that a sufficient number of MELK data sets are available in the form of fluorescence images.

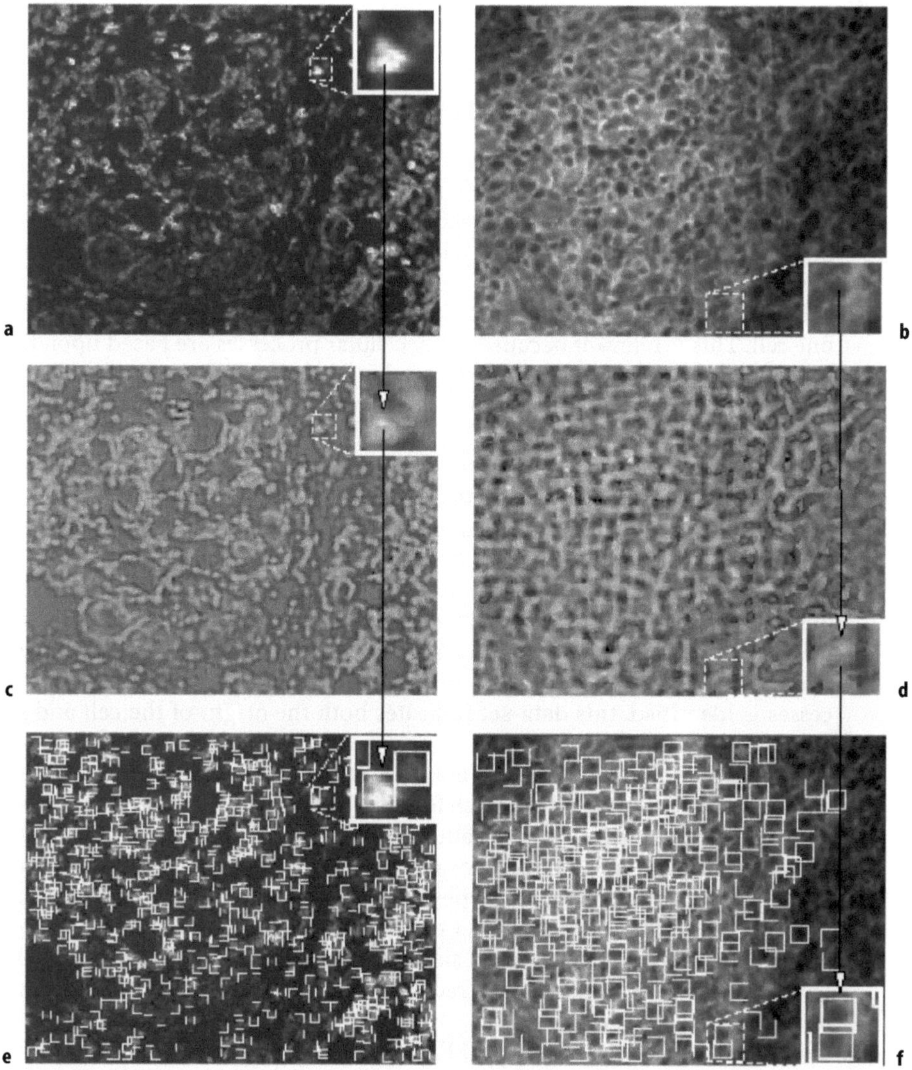

Fig. 7a–f. Three steps of cell detection for lymphocytes in muscle tissue and tonsil tissue are illustrated: The *first row* shows the two input recorded by our standardized technique of: **a** muscle invasive lymphocytes; **b** lymphocytes forming dense clusters within the tonsil. In both experiments the cells were immunolabeled against CD4. In the *second row* the evidence maps for: **c** muscle invasive cell; **d** clustering cells in the tonsil are presented. The evidence values $C(x) \in [0; 1]$, calculated by our neural classifier were mapped to *grey values*. A high value stands for high evidence for a fluorescent cell. The *large white boxes* in the corners show enlarged regions of each image and evidence map respectively. Following the *arrows* one can observe that high evidences correspond to positions of fluorescent cells. Local maximum analysis and thresholding delivers the final cell position within white frames, for both: **e** muscle tissue; **f** tonsil

8
Conclusions

The topological MELK proteomic approach has multiple implications in functional proteomics:

1. MELK allows one to address protein networks on the single cell level at the cellular and the subcellular level. It leads to the detection of functional linkages of proteins, which escape detection in large scale protein profiling techniques based on cell homogenates.
2. MELK allows one to detect these linkages in very small samples, down to only one cell. This is relevant because many cellular processes are based upon the specific actions of one or few cells. Examples are: (a) invasive tumour cells during minimal residual disease; (b) invasive pacemaker T lymphocytes in autoimmune disorders, such as polymyositis [15]; (c) single endothelial cells transdifferentiating to a muscle stem cell giving rise to an army of regenerating myogenic cells [2], and so forth.
3. Since cellular protein networks can be detected, which are specific signatures of different time-points during a dynamic change of a cell (biological states during cell polarisation or migration etc.), the corresponding MELK data sets are predictive concerning the cell fate, i.e. transdifferentiation of endothelial cells [2], translocation of invasive T lymphocytes in polymyositis [15]: once a complete MELK data set specific for different time points during such processes is identified, this data-set indicates both the origin of the cell and its destination.
4. The MELK-tag approach by using antibody search libraries for subcellular reference proteins allows one to search for disease-specific CPP-motifs on the basis of data-matching (normal vs pathologic tissue or cells). Thereby previously unknown disease-specific linkages of proteins can be assigned both to one specific cell type, multiple cell-types, or individual biological states of a cell type in the target sites of disease, i.e. small diagnostic biopsy samples. Since this approach allows one to detect simultaneously many different cell types in situ, it is possible to construct characteristic MELK-fingerprints of whole cellular systems.
5. On the basis of characteristic MELK-tag data sets a directed in-depth-search "around" the tag can be performed using large, extended antibody libraries derived from protein-specific sequence data provided by genomic information. Finally this MELK-tag search leads to the detection of lead proteins on the basis of disease-specific CPP-motifs. Lead proteins are considered to be first order target proteins, because they are specifically linked to disorders.
6. MELK provides a new platform to interlock target identification/prioritisation/validation with the industrialized drug discovery process. Particular interest is focused on the possibility to select hits or drug leads on the basis of their ability to shift an abnormal protein network to normal. This is a means by which hits or drug leads binding to a target without influencing cellular dysfunction can be selected out at an early time point in the drug devel-

opment process. Furthermore, hits or drug leads may be identified by MELK, which generate non-physiologic new protein networks likely to indicate unwanted side effects. Similarly, MELK can be used to screen patients' cells during clinical trials to identify drug effects on the cellular level and to correlate newly occurring non-physiological CPP-motifs with corresponding clinical data during treatment. The latter approach may raise the possibility that side effect targets could be identified in specific groups of patients, which would open new avenues for therapeutic intervention.

Finally, it will be interesting to link MELK with genomic approaches to cell function. For example, it might be examined if and how genomic mutations specifically influence functional protein networks in a cell, denoting the genes which are specifically interrelated.

9 Glossary

Toponome
Complete topological organization of the major molecule classes in a cell, as revealed by simultaneous mapping on the single cell level. It is the synopsis of all molecular networks in a cell, comprising networks of proteins, carbohydrates, lipids and nucleic acids.

Functional protein networks (FPN)
Networks of proteins, which are specifically linked to cell type, biological state of a cell or disease. FPN are detected by topological proteomic approach using MELK on the single cell level.

MELK technology
Multi-Epitope-Ligand-Karthographie (syn. whole cell fingerprinting), a high-throughput technology exploring the toponome.

s-CPP
Single combinatorial protein pattern, as revealed by MELK. The smallest elementary unit of a toponome, related to the topological protein complement of a cell.

CPP-motif
A specific set of s-CPP, which is characteristic for an individual biological state or function.

Topological proteomics
Research field of functional proteomics detecting functional protein networks on the single cell level, i. e. by using the MELK technology.

Lead proteins (LP)
Proteins, which are present in all s-CPPs of a CPP-motif. LP show the strongest linkage to the CPP-motif. In case of abnormal CPP-motif in disease, LP are considered to be first-order target proteins.

10 References

1. Schubert W (2000) Automated determining and measuring device and method. US Patent 6 150 173
2. Schubert W (1992) Antigenic determinants of T lymphocyte α/β receptor and other leukocyte surface proteins as differential markers of skeletal muscle regeneration: detection of spatially and timely restricted patterns by MAM microscopy. Eur J Cell Biol 58:395
3. Schubert W, Kontozis L, Sticker G, Schwan H, Haraldsen G, Jerusalem F (1988) Immunofluorescent evidence for presence of interleukin – 1 in normal and diseased human skeletal muscle. Muscle Nerve 11:890
4. Zimmermann K, Herget T, Salbaum J, Schubert W, Hilbich C, Cramer M, Masters CL, Multhaup G, Kang J, Lemaire HG, Beyreuther K, Starzinski-Powitz A (1988) Localization of the putative precursor of Alzheimer's disease – specific amyloid at nuclear envelopes of adult human muscle. EMBO J 7:367
5. Schubert W, Zimmermann K, Cramer M, Starzinski-Powitz A (1989) Lymphocyte antigen Leu19 as a molecular marker of regeneration in human skeletal muscle. Proc Natl Acad Sci USA 86:307
6. Schubert W, Prior R, Weidemann A, Dircksen H, Multhaup G, Masters CL, Beyreuther K (1991) Localization of Alzheimer βA4 amyloid precursor protein at central and peripheral synaptic sites. Brain Res 563:184
7. Schubert W (1991) Triple immunofluorescence confocal laser scanning microscopy: spatial correlation of novel cellular differentiation markers in human muscle biopsies. Eur J Cell Biol 55:272
8. Schubert W, Masters CL, Beyreuther K (1993) APP^+ T lymphocytes selectively sorted to endomysial tubes in polymyositis displace NCAM-expressing muscle fibers. Eur J Cell Biol 62:333
9. Schubert W, Schwan H (1995) Detection by 4-parameter microscopic imaging and increase of rare mononuclear blood leukocyte types expressing the FcγRIII receptor for immunoglobulin G in human sporadic amyotrophic lateral sclerosis (ALS). Neurosci Lett 198:29
10. Lendeckel U, Wex T, Ittenson A, Arndt M, Frank K, Maiboroda O, Schubert W, Ansorge S (1997) Rapid mitogen-induced aminopeptidase N surface expression in human T cells is dominated by mechanisms independent of de novo protein biosynthesis. Immunobiology 197:55
11. Schubert W, Agha-Amiri K, Mayboroda O, Rethfeldt C (1997) Dipeptidyl peptidase IV (CD26) and Alzheimer amyloid protein precursor (APP) in polymyositis. Adv Exp Med Biol 421:273
12. Kuznetsov AV, Mayboroda O, Kunz D, Winkler K, Schubert W, Kunz WS (1998) Functional imaging of mitochondria in saponin-permeabilized mice skeletal muscle fibers. J Cell Biol 140:1091
13. Vielhaber S, Kunz D, Winkler K, Wiedemann RF, Kirches E, Feistner H, Heinze HJ, Elger E, Schubert W, Kunz S (2000) Mitochondrial DNA abnormalities in skeletal muscle of patients with amyotrophic lateral sclerosis. Brain 123:1339
14. Haars R, Schneider A, Bode M, Schubert W (2000) Secretion and differential localization of the proteolytic cleavage products Aβ40 and Aβ42 of the Alzheimer amyloid precursor protein in human fetal myogenic cells. Eur J Cell Biol 79:400
15. Schubert W (2002) Polymositis, topological proteomics technology and paradigm for cell invasion dynamics. J Theoret Med 4:75
16. Bischoff R (1994) The satellite cell and muscle regeneration. In: Engel AG, Francini-Armstrong C (eds) Myology. McGraw Hill, New York, p 97
17. De Angelis LD, Berghella L, Coletta M, Lattanzi L, Zanchi M, Gabriella M, Cusella-De Angelis, Ponzetto C, Cossu G (1999) Skeletal myogenic progenitors originating from embryonic dorsal aorta coexpress endothelial and myogenic markers and contribute to muscle growth and regeneration. J Cell Biol 147:869

18. Schubert W, Friedenberger M, Haars R, Nattkemper T, Ritter H (2002) Automatic recognition of muscle invasive T lymphocytes expressing dipeptidyl-peptidase IV (CD26), and analysis of the associated cell surface phenotypes. J Theoret Med 4:67
19. Nattkemper T, Ritter H, Schubert W (2001) A neural classifier enabling high-throughput topological analysis of lymphocytes in tissue sections. IEEE Trans Inf Techn Biomed 5:138
20. Nattkemper T, Ritter H, Schubert W (1999) Extracting patterns of lymphocyte fluorescence from digital microscope images. Intelligent Data Analysis in Medicine and Pharmacology. Proceedings of the annual symposium of the American Medical Informatics Association. Washington, IDAMAP, p 79
21. Nattkemper TW, Wersing H, Schubert W, Ritter H (2000) A neural network architecture for automatic segmentation of fluorescence micrographs. Proceedings of the European Symposium on Artificial Neural Networks. Bruges, ESANN, p 177
22. Nattkemper T, Wersing H, Schubert W, Ritter H (2000) Fluorescence micrograph segmentation by Gestalt-Based Feature Binding. Proceedings of the IEEE-INNS-ENNS International Joint Meeting Conference on Artificial Neural Networks. Como, IJCNN
23. Nattkemper TW, Wersing H, Ritter H, Schubert W (2000) Automatic evaluation of multiparameter fluorescence micrographs with neural network architecture. In: Valafar F (ed) Proceedings of the International Conference on Mathematics and Engineering Techniques in Medicine and Biological Sciences. CSREA press, METMBS, Las Vegas, p 739
24. Hermann T, Nattkemper TW, Ritter H, Schubert W (2000) Sonification of multi-channel image data. In: Valafar F (ed) Proceedings of the International Conference on Mathematics and Engineering Techniques in Medicine and Biological Sciences. CSREA press, METMBS, Las Vegas, p 745
25. Mardia KV, Wei Quian, Shah D, de Souza KMA (1997) An algorithm for dividing clusters fluorescent stained nuclei. IEEE Trans Pattern Anal Mach Intell 19:1035
26. Dow AI, Shafer SA, Kirkwood JM, Mascari RA, Waggoner AS (1996) Automatic multiparameter fluorescence imaging for determining lymphocyte phenotype and activation status in melanoma tissue sections. Cytometry 25:71
27. Hanahara K, Hiyane M (1990) A circle-detection algorithm simulating wave propagation. Mach Vision Appl 3:97
28. Galbraith W, Wagner MCE, Chao J, Abaza M, Ernst LA, Nederlof MA, Hartsock RJ, Taylor DL, Waggoner AS (1991) Imaging cytometry by multiparameter fluorescence. Cytometry 12:579
29. Gerig G, Klein F (1986) Fast contour identification through efficient Hough transform and simplified interpretation strategy. Proc Int Conf Pattern Recognition 8:498
30. Ballard DH (1981) Generalizing the Hough transform to detect arbitrary shapes. Pattern Recognition 13:111

Received: April 2002

Author Index Volumes 51 – 83

Author Index Volumes 1 – 50 see Volume 50

Subject Index

MIX
Papier aus verantwortungsvollen Quellen
Paper from responsible sources
FSC® C105338

If you have any concerns about our products,
you can contact us on
ProductSafety@springernature.com

In case Publisher is established outside the EU,
the EU authorized representative is:
Springer Nature Customer Service Center GmbH
Europaplatz 3, 69115 Heidelberg, Germany

Printed by Libri Plureos GmbH
in Hamburg, Germany